ODD COUPLES

ODD COUPLES

Extraordinary

Differences

between

the Sexes

in the

Animal Kingdom

DAPHNE J. FAIRBAIRN

PRINCETON UNIVERSITY PRESS

Princeton and Oxford

FOR ALL THE ANIMALS

CONTENTS

ODD COUPLES

CHAPTER 1

Introduction

I spent my earliest days as a professional biologist happily tramping through the forests of southern Canada recording the survival, reproduction, and movement patterns of tiny mice that make their living among the detritus of the forest floor. My study subjects were the big-eared, dark-eyed deer mice that so often plague country cabins, apparently preferring the security and constant availability of food in human habitations to the perils of life in the wild. The official motivation for my study was to determine what causes the numbers of mice in a given population to fluctuate over time, and this, in turn, was motivated by the general question of what determines the numbers of individuals in any animal species. This was (and remains) a laudable goal, and I set forth on the study with full confidence in its scholarly legitimacy. Nevertheless, like most field biologists, I had chosen this type of research partly because it allowed me to spend much of my time out in the natural world rather than standing at a laboratory bench or hunched over a microscope. I had always been fascinated with the lives of wild creatures, and spending my time in the woods discovering how mice get through life seemed like just the ticket.

I captured my subjects in small metal boxes that my assistants and I placed strategically along mouse highways or close to mouse burrows. To entice the mice to enter our boxes, each was packed with a handful of nutritious seeds, a large dab of peanut butter, and a generous wad of cotton batting that the occupant invariably fluffed into a warm nest. The boxes were fitted with trap doors that closed when a mouse entered, and so we were able to catch the mice alive and release them again unharmed. The first time we caught a given mouse we would put a numbered, metal tag in its ear (much like a personalized ear stud) so that we could identify it when we caught it again. Our boxes were mouse magnets, and we caught hundreds of individuals, following many through most of their lives. To find out more about their individual personalities, I also brought each mouse into the laboratory for several nights (these are nocturnal mice) and ran it through a series of behavioral tests before taking it back to its capture site in the wild. Through all of this I got to know many of my mice quite well, and I was always delighted when old "friends" jumped out of their trap boxes in the cool of a morning. Once a mouse had finally disappeared from my study area, I would go back through my trapping records and reconstruct its life history. I could see where it had probably been born, where it traveled during its life, and how long it lived. If the mouse was a female I could tell when she was ready to mate, when and how often she had been pregnant and gave birth, and for how long she nursed her litters. My reproductive information was sketchier for males, but I could at least tell when they were ready and eager to mate.

Although my study was ostensibly designed to predict population numbers, as I got to know my mice I became increasingly absorbed by the obvious differences among them—and particularly by the different ways that males and females functioned in the population. For example, my females faced many hazards associated with pregnancy and lactation, and many died or lost their litters due to cold, rainy weather or lack of food. In contrast, my males ignored their offspring and spent the reproductive season searching for receptive females and interacting aggressively with other males. Their greatest hazard seemed to be displacement by more aggressive rivals. Although the numbers of both sexes showed similar seasonal fluctuations, it became clear to me that this demographic similarity masked profound and fascinating differences between the sexes. As it turns out, this revelation set the course for much of my future research and ultimately provided the motivation for this book.[1]

I've come a long way since my deer mouse days. I now ask questions about the evolution and adaptive significance of animal characteristics rather than about population sizes and demography, and several years as a fisheries biologist followed by several decades studying various insects and spiders have expanded my animal horizons beyond the familiar world of temperate mammals. Nevertheless, the differences between males and females that emerged so strongly in my studies of deer mice have remained a persistent theme. Time and again I have discovered how different life is for males and females in the natural world. The sex of animals affects their morphology, life history, behavior, and ecology, and this is true throughout the animal kingdom. The modest differences I observed in deer mice are trumped many times over by much more sexually dimorphic species in many different animal lineages. In some of these species the sexes differ so much that they would never be recognized as belonging to the same species had they not been observed actually mating or emerging from the same batch of eggs. In this book I ask why sexual differences are such a pervasive and significant part of the fabric of animal variation and, in particular, why males and females have come to differ to truly extraordinary degrees in some animal lineages.

I answer these questions by looking at sexual differences across the animal kingdom with emphasis on species where the differences are at the extremes of the distribution. These extraordinary species are particularly apt examples because they so clearly illustrate the division of reproductive function between the sexes and the extremes to which each sex can go to perfect its role. At one extreme are species in which males are large, muscular despots defending harems of much smaller females against constant challenges from rival males, whereas at the other extreme are species in which the females are solitary, large, and fierce, while their mates are minuscule sexual parasites capable of little more than producing sperm. My aim is to reveal what life is like for males and females at both ends of this continuum and to explain why such extreme differences have evolved.

As an evolutionary biologist I take an overtly Darwinian approach to understanding sexual differences. The basic assumption of this approach is that sexual differences are adaptations that increase the success of males and females in their respective reproductive roles. The currency of evolutionary adaptation is Darwinian fitness, which in this context means the number of descendants produced or, alternatively, the number of genes passed on to future generations. Thus, when I ask why male and female animals differ, I am really asking how the Darwinian fitness of males and females is enhanced by these sexual differences. Survival, particularly survival to reproductive maturity, is an important component of fitness in both sexes, but it is not the only determinant of success in leaving descendants. The number and quality of eggs or offspring produced are key components of fitness for females, and the number and quality of eggs fertilized or offspring sired are analogous components for males. These major fitness components can be further subdivided to measure more specific aspects such as survival through a particularly difficult part of the life cycle, escape from particular predators, or success in capturing certain types of prey. By definition, if traits are adaptive, they must enhance at least some of these aspects of fitness. This is the signal that the trait or at least certain values of the trait are being favored by natural selection.[2] The implication of relevance here is that we expect the traits typical of males to enhance male fitness, whereas those typical of females should enhance female fitness. In other words we expect the trait distributions typical of each sex to be adaptive for that sex.

In studies of sexual differences one type of selection has received particular attention, and that is selection that acts through differential success in acquiring mates. Darwin coined the term sexual selection for this type of selection, and he described its importance in the evolution of secondary sexual traits, particularly in males.[3,4] He presented copious evidence that males in many species are adapted for success in competition for mating opportunities and that many exaggerated male traits such as elaborate plumage, robust horns and antlers, and courtship songs and behaviors can be understood as consequences of sexual selection. Conversely, he argued that sexual selection was likely to be much less important in females, although he speculated that mate choice by males may have influenced the evolution of secondary sexual traits in human females.

Over the more than 140 years since Darwin published his extensive discussion of the evolution of sexual differences in *The Descent of Man and Selection in Relation to Sex,* legions of behavioral ecologists and evolutionary biologists have added to his evidence and confirmed the importance of sex-specific patterns of selection in producing sexual differences. The concept of sexual selection has now been extended to include competition among males for fertilization success during and after mating (for example, through courtship during copulation, competition among the sperm of different males in the female's reproductive tract, and preferential use of sperm by females). There is also increasing evidence that sexual selection on females is more significant than Darwin supposed.[5] However, these extensions are merely embellishments of Darwin's initial insights. His thesis that sexual differences are adaptive and caused by selection for different suites of traits in males and females is now solidly supported by volumes of theoretical and empirical studies,[6] and it forms the underlying theme of my explorations in the following chapters.

With this Darwinian perspective, I ask why male and female animals so often differ in morphological traits such as body size, shape, and color; in behavioral traits such as aggressiveness and patterns of migration or dispersal; in ecological traits such as what they eat and where they spend their time; and in life history traits such as the age at which they mature and the number of mates they acquire. In the next chapter I answer this question in a general sense by describing the fundamentals of male and female reproductive roles across the animal kingdom. This overview sets the context for chapters 3 through 10, which examine the lives of males and females in species with truly extreme sexual differences. My examples range from elephant seals and shell-carrying cichlid fish, where large and aggressive males defend harems of much smaller females, to deep-sea anglerfishes (called seadevils), in which females are large, fierce predators, and males live as tiny parasites permanently attached to the bellies of their mates. I describe orb-web spiders in which tiny males die spontaneously while *in copula* with their giant mates and open-ocean octopuses in which minuscule males ride on floating jellies as they search for females 40,000 times their size. My most extreme examples are species of marine tubeworms and burrowing barnacles in which adult females live sedentary lives firmly attached to a substrate while harems of dwarf males live permanently in or on their bodies. In each chapter I ask why the sexes are so disparate and how such gulfs between male and female function can possibly be beneficial. In chapter 11, I take a broader view and examine patterns of sexual differences across the entire animal kingdom. I ask why these patterns exist. For example, why can we almost always distinguish males and females by differences in size and shape? Why are females usually the larger sex, and why are species with giant females and dwarf males so much more common than the reverse? Why are color differences restricted to relatively few animal groups and why, when they do occur, are males likely to be brightly colored while females are dull or camouflaged? In all of these chapters my focus extends beyond sexual selection, animal sex, and why males so often have ornaments and weapons not found in females (although all of these themes are certainly included). Through my descriptions of extraordinary species and my survey of variation across the animal kingdom, I explore what it means to be male or female in the broadest sense and illustrate, as vividly as possible, the impact that this division of reproductive function can have on virtually all aspects of an animal's life.

Chapters 2 and 11 provide important background and context for my exploration of sexual differences, but the heart of the book is the descriptions of the eight species (or in some cases, groups of very closely related species) highlighted in chapters 3 through 10. Each provides a unique and excellent example of widely disparate sex roles, and together they offer a grand overview of extreme sexual differences in animals. However, these few examples are by no means an exhaustive sample. Truly extreme sexual differences, where the two sexes would scarcely be recognized as belonging to the same species, are found in many different species distributed across at least eleven animal phyla.[7] Selecting eight examples from this diverse array of candidates was a daunting task. Although I admit to a modicum of serendipity and even a dash of subjectivity in the selection process (I just found some animals more fascinating than others), I did use a set of objective criteria to narrow the field. My foremost criterion was that the sexes had to show extreme differences in body mass. Because mass is a property of all animals, I could use the ratio of the masses of adult males and females as an objective criterion for comparing the magnitude of sexual dimorphism[8] among species as disparate as elephant seals and octopuses. Mass also correlates strongly with many other aspects of the biology and ecology of animals, including aspects of physiology (metabolic rate, production and dissipation of heat, energetic costs of movement), morphology (robustness of supporting skeletons, relative size of horns and antlers), performance (maximum speed, acceleration), life history (age at maturity, lifespan, clutch or brood size), and ecology (home range size, dispersal distance, population density).[9] Thus, pronounced sexual dimorphism in body mass (sexual size dimorphism) invariably correlates with sexual differences in many other traits.[10] Conversely, marked sexual differences in life history, behavior, ecology, or morphology seldom occur in the absence of sexual size dimorphism. Extreme sexual size dimorphism thus provides an objective, quantitative, and universal indicator of similarly extreme differences between males and females in many, if not most, aspects of their lives.

Having narrowed my choices to species with the greatest magnitude of sexual size dimorphism, I next limited my choices to species in which the ecology, life history, and behavior of both sexes were well described in the scientific literature. Where possible I chose species in which both sexes had been followed from birth to death, so that I could trace the entire life cycle of each sex and relate differences in early developmental trajectories to the disparate adult characteristics.

My final requirement was that the selection of examples should represent as much animal diversity as possible. With room for only eight examples, that was a tall order. Even though fewer than 4 percent of living animal species are vertebrates, we know far more about sexual differences in vertebrate species than in the vast assortment of invertebrates, and I could easily have populated my chapters entirely with vertebrate examples. As it is, I include a mammal, a bird, and two fishes, and the other four example species represent only three of the thirty invertebrate phyla: the mollusks, annelid worms, and arthropods. Given that the latter phylum contains more than 78 percent of living animal species, it seemed appropriate to include examples from two classes: spiders from the class Arachnida and barnacles from the class Maxillopoda. Although these eight examples cannot possibly encompass the full diversity of sexual differences, they do represent the extremes at both ends of the continuum—from a species where males average almost 13 times heavier than females to one where females can be as much as 500,000 times heavier than males.

The book offers much information that will be of interest to biological scholars with expert knowledge of at least some of the concepts and biology described. However, it is my hope that the descriptions of extraordinary animals and summaries of patterns of sexual differences across the animal kingdom will also be read and enjoyed by readers who are not biological scholars but are simply interested in the diversity of animal life or in the biological roots of our own sexual differences. (I discuss sexual differences in humans only briefly at the end of the book, but the principles and overall trends that I describe apply no less to us than to any other animal species.) To serve potential readers with such disparate backgrounds and interests, I provide notes at the back of the book signified by superscripts in the text. These notes contain scientific citations supporting various statements in the text, sources for technical data, and additional explanations, details, or caveats that extend the information in the main text. In deference to nonspecialist readers I have avoided the use of biological jargon as much as possible. The terms that I have used are shown in italics when they first appear and are listed in a glossary at the back of the book for easy reference. To maintain the flow of the narrative and to make the book more enjoyable for nonspecialists, I use common names rather than scientific names for all but the major players in my chapters. The corresponding scientific names can be found in appendix A, listed alphabetically by common name. I also list the major animal lineages (phyla with included classes) in alphabetical order in my tables, rather than clustering them according to their degree of evolutionary relatedness. This allows readers to easily find a given phylum or class, even if they do not know the evolutionary history of that taxon. I offer my apologies to professional colleagues who may be uncomfortable with these breaches of standard scholarly protocol.

CHAPTER 2

The Roots of Sexual Differences

Why Male and Female Animals Differ

To understand why having two sexes seems to work so well for animals, we need to see this curious sexual dichotomy for what it is, a division of reproductive effort into separate male and female roles. A few animals are able to reproduce without sex, and some are *hermaphrodites*, combining male and female functions in the same individual, but the vast majority of animal species divide their reproductive function into distinct male and female sexes. This allocation of reproductive function, called *dioecy*, predominates in animals as disparate as mammals, insects, roundworms, and clams and is clearly the prevalent pattern of reproductive allocation for most animal lifestyles. It occurs in twenty-six of the thirty-one phyla and is the dominant or sole strategy in seventeen phyla, including the Arthropoda (by far the largest, containing insects, spiders, crustaceans, and relatives) and the Chordata, the phylum to which we belong.[1] By comparison, the five phyla that lack dioecy are all very small and together comprise less than 0.16 percent of living animal species. This is admittedly a crude measure of the prevalence of dioecy, but it is sufficient to demonstrate that having separate sexes is the predominant reproductive strategy in most animal groups. Put succinctly, the vast majority of animals spend at least the adult phase of their lives as either males or females.

Surprisingly, although dioecy is characteristic of most animal species, the biological mechanisms that determine sex are highly variable.[2] As an initial example, consider how sex is determined in humans. Our sexual destiny begins at the moment of conception, when the sperm and egg fuse to form the first cell of a new human being. The genes that will influence the sexuality of the new individual are scattered throughout its genome on many of its twenty-three chromosome pairs, but one pair, appropriately called the sex chromosomes, holds the key to sex determination. The chromosomes comprising this pair come in two forms, designated X and Y. Women have two X chromosomes, but males have an X and a Y chromosome. This means that our biological sex is determined by which sex chromosome we inherit from our fathers. If the fertilizing sperm carries a Y chromosome, the embryo will begin to differentiate as a male about seven weeks after fertilization when a gene on that chromosome (called *Sry*) initiates a biochemical chain of events that causes development of testicular tissue. In the absence of the *Sry* gene, ovarian tissue differentiates at about twelve weeks post fertilization, and the fetus becomes female.[3]

Sex chromosomal systems that appear to be analogous to our XX/XY system occur in many other animal lineages, including many worms,[4] insects, crustaceans, and most vertebrates other than birds. However, it is clear that Y chromosomes are not always essential because some species in these groups lack Y-chromosomes entirely. In these systems, designated XX/XO, sex is determined by the number of X-chromosomes, with two X chromosomes designating females and one designating males. In other lineages the chromosomal system is reversed so that males are produced when the two sex chromosomes match and females when they do not. This is called the ZZ/ZW system and is found in most birds, butterflies, snails and slugs, and occasionally in fishes and reptiles. As with the Y chromosome, the W chromosome is sometimes missing so that females have only one Z chromosome, whereas males have two.[5] Some animals lack sex chromosomes altogether and instead use other genetic cues to determine sex. For example, in ants, bees, wasps, and some beetles and rotifers, unfertilized eggs develop into males, whereas fertilized eggs develop into females.[6] There are even species in which sex is determined by a number of different genes distributed on various chromosomes.[7]

Even more surprising than this diversity of genetic mechanisms is the observation that some of the most extreme sexual differences occur in species with no obvious genetic differences between the sexes. In these species sex-specific developmental trajectories are apparently turned on by environmental or social cues at critical stages of development.[8] A bottom-dwelling marine worm called the green spoon worm is a good example of this. The sex of green spoon worms is determined by the substrate on which the larva settles when it is ready to mature. If a larva settles on an empty patch of ocean bottom, it metamorphoses into a female and grows to become a large, sac-like, sedentary worm with a long, thin proboscis with which it captures food. If, on the other hand, the larva happens to land on a female, it almost always metamorphoses into a tiny male, moves down the female's proboscis, and sets up shop inside her uterus where it devotes the rest of its short life to producing sperm. Amazingly, male and female spoon worms manage to develop one of the most extreme sexual dimorphisms in the animal kingdom from the same set of genes.

From this brief survey it is obvious that animals have evolved many different ways to produce males and females. Some have a set of genes that always occur in one sex but not the other, but in others males and females develop from identical genomes. If there is a universal key to sexual differentiation it is not sex-specific genes but, rather, sex-specific patterns of gene expression.[9] I like to envision the development of the suites of male and female traits from the same genotype as analogous to a pianist playing two different sonatas on the same piano. The same set of keys (genes) is present in both cases but they are played (expressed) at different times and for different durations in the two sonatas. The result can be two remarkably different pieces (suites of male and female traits). The initial trigger that determines sex simply opens one or the other songbook and perhaps plays the first few notes. Even in species that have sex chromosomes, genes on the sex chromosomes initiate the cascade of developmental processes leading to sexual differentiation, but the responding genes are distributed throughout the genome. The resulting sex-specific patterns of gene expression occur throughout the body in both reproductive and nonreproductive organs and tissues and produce fundamental differences between the sexes in almost all aspects of their biology, from biochemistry and physiology to morphology and behavior. Sex-specific timing of gene expression also regulates the timing of life events, producing differences between the sexes in growth trajectories, age at maturity, and even lifespan.

Although sex-specific patterns of gene expression may seem like a straightforward answer to the problem of producing two distinct suites of traits from a single genome, the mechanisms by which these patterns are orchestrated remain somewhat of a mystery. It is clear that animals are much more than simply the products of the basic genes inherited from their parents. In each new individual these genes interact with binding proteins called histones and regulatory factors such as hormones that influence their expression patterns. In some cases the expression of a gene may even depend on which parent it comes from. For example, only the copy inherited from the mother might be expressed, although both maternal and paternal versions are passed down to offspring. Elucidating effects such as these (called *epigenetic effects*) and discovering how they produce integrated and adaptive suites of traits are among the most exciting areas of research in the field of evolutionary genetics, and we have much to learn about how these effects are orchestrated in different animal lineages.[10]

Having established that dioecy is the prevalent theme among animal species, we are still left with the question of *why* this should be so. What is so special about having males and females? If all animals produced males and females the same way, we might surmise that dioecy is prevalent simply because all living animals have inherited this developmental program from a common, dioecious ancestral species, and we are basically stuck with it. Although it is true that the ancestor of all animals was probably dioecious,[11] it is also clear that the biological mechanism for determining sex has shifted many times during the subsequent evolutionary histories of modern phyla. This makes it very unlikely that living animals are dioecious simply because dioecy is a fixed, ancestral pattern. On the contrary the diversity of sex-determining mechanisms tells us that selection must be continually favoring specialization for male and female sex roles so that dioecy is maintained or reevolved even when the underlying biological determinants change. This conclusion brings us back to the fundamental question: Why does having two sexes seem to work so well for animals? To answer that question, we need to consider what each sex actually does and how the two sexes interact to pass their genes on to future generations.

The essence of sex in animals is that each sex produces germ cells or *gametes* that carry one copy of the parental chromosomes. By definition males produce smaller and more mobile gametes (sperm), and females produce larger, more nutrient-rich gametes (eggs) that are not capable of independent movement. New individuals with the normal double (*diploid*) chromosome complement are produced through fusion of egg and sperm nuclei within the egg. Accordingly, the fundamental reproductive function of most male animals is to produce large numbers of sperm that are specialized for finding and fertilizing eggs. Male Atlantic salmon and Atlantic cod, for example, release about 190 billion sperm per ejaculate, whereas human males produce 100–300 million. By releasing large numbers of sperm in one go, the males increase the probability that at least some sperm will successfully reach an egg. Large numbers of sperm are especially important in aquatic animals where males release their sperm into the water, and the ejaculate rapidly becomes diluted. If the spawning males are directly competing with other males to fertilize a given female's eggs, as in salmon or cod, the benefits of releasing massive quantities of sperm are even greater.[12]

One way of solving the dilution problem and reducing competition among ejaculates is for males to release their sperm within the female's reproductive tract so that her eggs can be fertilized before they are spawned or laid. To this end, mechanisms for internal fertilization have evolved in many different animal lineages and are now found in at least twenty-one phyla.[13] However, internal fertilization brings challenges of its own for males, including the need for a specialized copulatory organ and for a set of social interactions that permit males to secure copulations by persuasion, intimidation, or both. Further, although it greatly ameliorates the dilution problem, internal fertilization seldom completely frees males from sperm competition. In many species females mate with more than one male and are able to store the sperm from multiple mates, often for large portions of their reproductive lifespans. In these species sperm competition can occur within the female's reproductive tract, and just as in free-spawning species, males benefit from releasing voluminous ejaculates containing large numbers of sperm.[14]

Competition among sperm from different males affects more than just numbers of sperm. In some species it has promoted the evolution of complex sperm morphologies, very large sperm, or even sperm dimorphisms where one sperm morph is specialized for displacing the sperm of rival males while the other fertilizes the eggs. Males have also evolved numerous secondary ways of giving their sperm the edge. They may use their intromittent organ to physically remove or destroy leftover sperm from previous males (some male damselflies and beetles even have specialized scoops or brushes to do this job). They may also incorporate chemicals in their seminal fluid that alter female behavior or physiology to increase the probability that their sperm will fertilize her eggs. For example, tiny male fruit flies are notorious in scientific circles for their ability to chemically manipulate females, causing them to delay mating with other males while at the same time prompting them to lay more eggs. The net effect of this devious tactic is to coerce the female into fertilizing her eggs with the sperm of her present mate before mating again. Still other males have evolved ways of stroking or otherwise stimulating the female during copulation to encourage her to preferentially store and use their sperm to fertilize her eggs.[15]

In their quest to mate with females and fertilize eggs, males also compete with each other in many more obvious ways. We are all familiar with species in which males attempt to physically displace rivals or defend breeding females from other males. The typical scenario is for males to threaten and physically challenge each other in defense of a single, highly fecund female (as in orb-web spiders) or, more commonly, in defense of a group of reproductive females (as in lions, bighorn sheep, and elephant seals). When males compete directly in this manner, those that are larger overall or have larger armaments or weapons often have the advantage. As a consequence males in such species are often larger than females and possess various sexual armaments. Familiar examples of the latter include antlers and enlarged canine teeth in some male mammals; long, hooked jaws in male salmon; horns of various sorts in insects; and leg spurs in roosters and other male gamebirds. Another very common male tactic is to focus on success in attracting females. In many species females actively choose their mates, and males compete by displaying themselves to females in hopes of being chosen. The type of display that males use depends to some extent on the medium in which they are displaying (i.e., in the air, on the ground, or in the water) and on the sensory capabilities of their females, but in all cases the display is tailored to attract female attention, often through multiple sensory modalities. For example, displaying male birds typically employ a combination of colorful plumage, loud vocalizations, and choreographed dances or flights. Male crabs, insects, fishes, and mammals often include chemosensory attractants along with visual and auditory displays. In species where males compete with each other in this indirect way by vying to attract females, they are often much showier than the females, which typically remain rather inconspicuous to avoid the attention of predators. Males in many species employ elements of both of these tactics and are likely to be both larger and more conspicuous than their mates. Barnyard roosters with their large size, long legs, sharp leg spurs, bright combs, showy plumage, aggressive temperament, and loud calls are a familiar example of this type of combined mating strategy and of the male characteristics that it produces.[16]

Several other male tactics for acquiring mates are less common but have nevertheless evolved independently in numerous animal groups. Males may attempt to entice females to mate with them by offering gifts of food or by providing secure sites for the eggs or offspring. Some males even help the female care for the offspring, or more rarely, they may take full responsibility for this themselves. If they do this they often rear the eggs of several females at once. For example, in fishes that spawn on lake or stream beds or on the shallow ocean bottom, it is not uncommon for males to build nests in which they entice several females to lay eggs.[17] The males then defend these nests against other males and predators, and they may even clean the eggs and fan the nest to provide well-oxygenated water. Sticklebacks and bluegill sunfish are two familiar examples of species that have adopted this tactic. Fish paternal care reaches its extreme in seahorses and pipefishes, many of which have a male abdominal brood patch or pouch where the eggs are deposited and nurtured until they hatch. Males have also taken over the role of sole parental providers in several species of shorebirds (jacanas, phalaropes, and sandpipers) and ratites (rheas, ostriches, emus, cassowaries). Males of several groups of tropical frogs guard their eggs, and in some species of poison dart frogs, the attentive males even carry newly hatched tadpoles on their backs, transporting them over land from small hatching pools to freshwater streams where they will develop into adults. Male parental care occurs in some insects as well, notably in many species of belostomid water bugs whose females lay their eggs on the backs of males that then carry the eggs around until they hatch. Sea spiders (which are not true spiders, but rather strange, long-legged animals in the arthropod class Pycnogonida) also delegate offspring care to the males, and the males often carry clutches of eggs from several females arranged along specialized appendages called ovigers.

Although paternal care as in the above examples occurs in a number of different animal groups, it is never a common strategy. In the vast majority of species the father's participation ends with spawning or copulation. Securing fertilizations is therefore the driving motivation for much of male morphology, life history, and reproductive behavior. I have already described three ways in which males typically secure fertilizations: defeating other males in contests for mates, being chosen to mate by discriminating females, and, once a mating has been achieved, ensuring the success of their sperm in competition with the sperm of other males. In many species males employ all of these tactics, albeit to varying degrees. (The examples in chapters 3, 4, and 5 nicely illustrate the interplay of male contest competition and female choice.) However, a much less-appreciated challenge for males is simply finding females when they are rare and widely dispersed. The risk of failing to find a mate is particularly high for species that live at low densities and do not form mating or social aggregations. (Species that, unfortunately, are not good subjects for detailed study.) Many species found in the open ocean or on the deep sea floor fit this description. Faced with little risk of competition from other males and considerable risk of failing to find a female, males in these species specialize in traits that increase their mate-searching capabilities. They tend to be small, highly mobile, and equipped with exaggerated sensory organs that allow them to sense females at a distance.[18] Since the females are also at risk of failing to find a mate, they tend to be specialized for long-distance male attraction, often using chemo-attractant *pheromones*, *bioluminescence* or both, and they are unlikely to be choosy when a male appears. As we shall see in chapters 6 through 10, these are the species in which sexual dimorphisms become most exaggerated, with males sometimes tens of thousands of times smaller than females and structurally so different that it can be difficult to assign males and females to the same species.

Whereas males focus on securing fertilizations, females use very different strategies to maximize their reproductive success. The fundamental female reproductive role is to produce well-provisioned eggs that provide nourishment and protection for the developing embryos. Because eggs are typically several orders of magnitude more massive than sperm, females make relatively few of them when compared to male sperm production. In humans, for example, eggs are 85,000 times more massive than sperm, and on average women produce fewer than 500 mature eggs over their lifetimes compared to the trillions of sperm produced by men.[19]

To appreciate the value of a well-provisioned egg, consider the chicken egg you may have had for breakfast.[20] Large chicken eggs with large and nutritious yolks are not only grade A for eating, they also produce the largest, healthiest chickens. Chick weight at hatching is determined primarily by egg weight at laying and is typically about 62 percent to 78 percent of egg weight. That means that chicks from large eggs are larger at hatching. Chicks from large eggs also grow faster after hatching, so the beneficial effect of egg size compounds as the chick grows. An increase of 1 g in egg weight results in a corresponding increase of 2–13 g in the weight of the chicken at six to eight weeks of age (when, sadly, they become "broilers" and their story ends). Researchers have documented similar positive effects of egg size on various measures of offspring quality (e.g., size, growth rate, survival, and physiological and behavioral performance) in a great diversity of animal species, including birds, lizards, amphibians, fish, insects, and sea urchins.

Given the positive effects of egg size on offspring quality, you might think that all females would produce enormous eggs, but this is clearly not the case. In spite of the advantages of large eggs, females in many animal species produce surprisingly small eggs. One reason for this is that egg size tends to be negatively correlated with the numbers of eggs produced in any given batch or clutch. Making larger, better-provisioned eggs therefore usually means making fewer eggs, at least within that reproductive episode. The range of female reproductive strategies across the animal kingdom strongly reflects this trade-off between egg size and egg number.[21] At one end of the spectrum are females that produce only one large egg at a time. Female New Zealand brown kiwis are the most extreme example of this strategy. They lay one enormous egg that weighs about 400 g (14 oz), which is about 25 percent of the female's body weight and proportionally the largest of all bird eggs.[22] Kiwis have been successful having only one offspring at a time because they evolved in an environment with few, if any, predators and because one or both parents incubate the huge egg for about eighty days and then protect the chick for about two weeks after hatching. By virtue of its size, benign environment, and the care of its parents, the single kiwi offspring has a pretty good chance of surviving to adulthood.

At the other end of the spectrum are females that produce huge batches of tiny eggs that they completely ignore. North Atlantic cod provide a good example of this strategy.[23] Female cod typically spawn 500,000–1,000,000 eggs, and each egg weighs only 1.6 mg (0.0016 g), about 0.00008 percent of the average female's body weight. The whole batch of eggs weighs about 25 percent of the female's weight; so assuming that the eggs have a similar energetic content, kiwis and cod are allocating about the same proportion of their resources to their eggs. They are just trading off egg numbers and egg size very differently. Cod need to produce large numbers of small eggs because the challenges facing their offspring differ greatly from those experienced by kiwi chicks. Cod spawn their eggs and sperm in the sea, and the fertilized eggs float up to the surface waters and drift in the plankton. As in other animals, the size at hatching, growth rate, foraging ability, and survival of larval cod all increase with egg size, and the eggs are smaller than they need to be to remain buoyant, so one would think that female cod should produce larger eggs. The reason they do not is that almost all their eggs or larvae are eaten before they mature so that very few of a female's million or so potential offspring ever become adults. For cod the successful strategy is to produce so many eggs that, even with such a low probability of survival, some offspring will successfully reach adulthood. For an animal to produce such large numbers of eggs, each egg has to be small.

Whether producing a few large or many small offspring, female animals typically devote a high proportion of their energy and resources to the task. One indication of how much they devote before birth is how much the clutch or litter weighs at birth as a proportion of the mother's mass.[24] In female fishes this averages about 12 percent and can be as high as 31 percent. By comparison, male fishes devote an average of only 3.7 percent of their body mass to reproductive tissues and sperm. In mammals litter mass ranges from less than 2 percent of mother's mass in the largest species to more than 33 percent in the smallest species, whereas in male mammals even the smallest species devote less than 2 percent of their body mass to testes and sperm. The overall champions among vertebrates are female snakes, with proportional clutch or litter masses averaging 31 percent and exceeding 60 percent in some species. However, even these impressive values are small in comparison to the proportional clutch masses for arthropods such as insects, spiders, water fleas, and crabs, which often exceed 50 percent and can be as high as 75 percent. By comparison, the animal champion among males is the tuberous bushcricket, with testes that weigh up to 14 percent of the total weight of adult males.[25]

For the vast majority of animal species the mother's attention to her offspring ends when she gives birth or lays her eggs. However, even females that abandon their eggs or neonates usually try to enhance the success of their offspring by depositing them in places and at times that favor survival and growth. For example, females that spawn in the marine environment do so when the water temperature, tides, and currents combine to favor egg fertilization and larval growth and maturation. Similarly, females that lay eggs on substrates generally seek sites that provide protection for the eggs and feeding opportunities for the newly hatched young. In many species of plant-feeding insects, for example, mothers spend considerable time searching for specific plants on which to lay their eggs. When an appropriate plant is located, the female then lays her eggs only on specific parts of the plant such as within a developing flower, close to a growing shoot tip, or on the protected undersides of leaves. In each case the female chooses the location suitable for the food requirements of her offspring. To accomplish this complex task females have evolved specialized sensory capabilities for detection of plant chemicals, and many also have specialized morphological adaptations for cutting into soft plant tissues or boring into wood to lay their eggs. Thus, even in species without overt maternal care, mothers are likely to have behavioral, morphological, and sensory specializations that enhance the success of their offspring after hatching or birth.

Maternal specializations are even more obvious in species where mothers continue to care for their offspring beyond birth or egg deposition. Mammals have taken postnatal maternal care to its extreme by providing specialized nutrition for their young in the form of milk from the mother's mammary glands. The energetic cost of this is huge. The daily energetic expenditure of lactating females typically increases by 100 percent to 200 percent over nonreproductive needs, and by 20 percent to 120 percent over the cost of the pregnancy itself.[26] Most mammalian mothers also cuddle their offspring to assist them in thermoregulation, protect them from predators, and help them make the transition from nursing to foraging for food. Rarely do they receive any assistance from the fathers, who are more likely to be a threat than a help (absent dads are the rule in more than 90 percent of mammal species, and males of many species will readily eat juvenile members of their own species).[27] Maternal care is also prolonged in most bird species, although male birds tend to be considerably more helpful than male mammals. In about 90 percent of bird species males participate in nest building, egg brooding, feeding the offspring, or guarding the nest or offspring, but rarely are they the primary or sole caregivers.[28] Birds that feed their young by bringing food to the nest increase their daily energy expenditure by an average of 30 percent to 50 percent, and this cost is generally higher for females because they do more than their share of feedings.[26] Although female birds lack the obvious maternal specialization of female mammals (mammary glands), they often have subdued coloration, seldom vocalize, and tend to remain hidden when eggs or chicks are present, all adaptations that reduce the risk of predation on the eggs or offspring.

Mammals and birds tend to dominate our view of animal characteristics, but with approximately 4,800 and 9,900 species respectively, they comprise only about 1 percent of living animal species. The emphasis on offspring care in these two, small vertebrate lineages gives a very biased view of what is typical of animals in general. Few other animals extend their care beyond egg laying or birth, and those that do tend to restrict themselves to guarding or brooding their eggs until they hatch.[29] For example, many female mollusks, barnacles, and water fleas brood their eggs within their mantle cavity, and female crabs, lobsters, shrimps, and copepods often carry their eggs on specialized abdominal appendages. In other invertebrate lineages, including some bryozoans and brittle stars, females brood their eggs in specialized brood pouches within their body cavity. In most cases the offspring leave their mother's care at or shortly after hatching, but, rarely, females continue to care for their young through the larval stages as well. The most well-known examples of this among invertebrates are the colonial termites and hymenopterans (ants, wasps, and bees) in which the larvae are tended and fed by cadres of sterile female workers until they metamorphose into adults.[30] This is all to say that female reproductive roles sometimes extend beyond egg laying to include protecting and sometimes providing nutrition for offspring until the offspring are able to fend for themselves. Although the same can be said for males in a few lineages, commitment to parental care beyond egg fertilization is primarily a female characteristic. It is a logical extension of the larger initial investment that females make in individual eggs.

From the preceding descriptions of male and female reproductive roles, it is not difficult to imagine why specialization for male or female function often results in differences between the sexes in many of their biological characteristics. In addition to looking and behaving differently males and females often grow at different rates, become sexually mature at different ages and sizes, have different expected lifespans, and spend much of their time in different places. These differences exist because the suite of traits that maximizes a female's ability to successfully produce offspring is seldom identical to the suite of traits that will achieve the same ends for a male. For example, females have to accommodate developing eggs or embryos within their bodies, and so they are often larger and thicker-bodied than males. Similarly, because they have to live long enough to produce eggs or live young whereas males only need live long enough to mate, females in many species live much longer after mating than males do. The yellow garden spiders, blanket octopuses, bone-eating worms, and burrowing barnacles that I describe in chapters 6, 7, 9, and 10 are all examples of animals that carry this pattern of sexual differences to the extreme, combining giant, long-lived females with tiny, short-lived males. The giant seadevils (a type of deep-sea anglerfish) described in chapter 8 follow much the same pattern, except that dwarf male seadevils manage to extend their lifespans by becoming parasites on their mates. The extraordinary sexual differences in these five examples seem truly bizarre not only because they are so extreme, but also because they are so different from our everyday experiences. We are more familiar with species in which males are the larger, more flamboyant sex, with obvious adaptations for attracting females, repelling rival males, or both. Although relatively uncommon among animals in general,[31] this is the prevalent pattern in most birds and mammals and the one we find in most of our domesticated animals. It is carried to its most extreme in the elephant seals, great bustards, and shell-carrying cichlid fish that I describe in chapters 3, 4, and 5. In these species males mature at later ages and larger sizes than their mates in part because sexual selection favors large and aggressive males. As we shall see, however, sexual selection on males is far from the whole story; the pattern of selection on females is equally important. The message that emerges from all eight of my examples is that within each species both sexes are uniquely adapted to fulfill their own reproductive roles, and only by understanding how life is experienced by each can we hope to understand the differences between them.

CHAPTER 3

Elephant Seals

Harems, Hierarchies, and Giant Males

My first encounter with elephant seals occurred on a gray day in early January of 2003 as I was driving down the coast highway in California with my husband, Derek, and daughter, Robin. We were alone on the highway and all enjoying the spectacular cliffs, ocean views, and generally wild country that stretches south from Monterey Bay to Santa Barbara. We came down a smooth sweep of road from the high cliffs of Big Sur to the gentler coastal hills north of Point Conception. The highway curved over to the very edge of the low beach cliffs, and suddenly, on the sand below the bluff, an elephant seal emerged out of the mist and drizzle. We quickly pulled over, staring in disbelief. Grabbing cameras and binoculars, we piled out of the car and found ourselves on the edge of a breeding colony of northern elephant seals. We spent several hours observing and photographing the huge males, smaller females, and tiny, newborn pups sprawled on the sand. We watched bulls humping up the beach only to be waylaid by resident bulls and chased unceremoniously back into the surf. We were amazed at the speed of the huge, lumbering males, and we cringed when the chases overran the helpless pups, at risk of crushing them. Several times the big males faced each other, making loud, burbling honks through their noses, swinging their great heads and necks, and angling to stab each other with their canine teeth. Most were battered and bleeding from such encounters, with pink blubber showing clearly through the open wounds on their chests and necks. Through all of this, the females lay helter-skelter and unperturbed on the sand, most with black pups at their sides. Occasionally a male would hump himself over to a female and demand attention by raising his head and honking though his great elastic nose. The female's response was invariably to croak back in annoyance at being disturbed. Always the tiny pups seemed in danger from the great, lumbering males, and we did see several get flattened under males as they courted females or chased other males. Fortunately, those pups all emerged alive, but there were a number of lifeless pups lying alone, deserted by their mothers, and we wondered if they had been casualties of such encounters.

In the long view of evolution, this unexpected colony of elephant seals breeding along the edge of a modern, paved highway in twenty-first century California seems almost miraculous. Northern elephant seals were hunted nearly to extinction in the 1880s, and only a relict population of no more than one hundred breeding animals remained on Guadalupe Island off the coast of Baja California. Fortunately, this population was protected, and rather than going extinct, which it might well have done given its small size, it grew and produced emigrants that founded new colonies. Over the past century the total population of northern elephant seals continued to grow and spread to recolonize much of its former range from Mexico to Canada The site we happened upon occupies a series of sandy coves just south of the Piedras Blancas lighthouse along the San Simeon coast of California.[1] The first colonists appeared at this site in 1990, and the first pups were born there in January of 1992. By the time we visited in 2003, there were approximately 10,000 seals in the population (called a rookery), with more than 2,000 new pups born, and the overall population of northern elephant seals had grown to over 150,000. I revisited the site in January of 2010, and by that time the rookery had grown to over 15,000 with more than 4,000 pups born each year (plate 1). The beach was even more crowded and chaotic than it had been in 2003, striking evidence of the spectacular recovery of these magnificent animals. The current world population of northern elephant seals is estimated to include approximately 170,000 animals, and the species is now listed as "of least concern" on the International Union for Conservation of Nature (IUCN) red list of threatened species.[2]

Elephant seals are in the family of true or earless seals (the family Phocidae), and they are sufficiently distinct from all other seal species to be placed in their own genus, *Mirounga*. Northern elephant seals are *Mirounga angustirostris* and their sister species, the southern elephant seal, is *M. leonina*. Southern elephant seals are the more abundant of the two species, with an estimated world population of about 650,000.[3] They have their rookeries on scattered subantarctic islands and a few mainland sites in southern Argentina. About half of the population breeds on the remote island of South Georgia in the south Atlantic, approximately 1,900 km (1,200 miles) east of Cape Horn at about 54° south latitude. The two species of elephant seals are the largest and most sexually dimorphic of the living pinnipeds (table 3.1, figure 3.1). Southern elephant seal males grow considerably larger than their northern cousins, but females of the two species are quite similar in size. As a result southern elephant seals are the unquestioned champions of size dimorphism not only among pinnipeds but among all mammals, with mature males typically weighing seven to eight times more than females (plate 2).[4] Aside from this marked difference in male body size, the two species are very similar, and most of what I will be telling you about their behavior, life history, and ecology applies equally to both (table 3.1).[5]

My two visits to the colony at Piedras Blancas occurred during the time of year when female elephant seals are giving birth and males are seeking mates. Although both sexes are present in the rookery at this time, they share little in common save the brief moments of copulation. From the female perspective the rookery provides a safe place for giving birth and nursing pups.[6] By the time females return to the rookery to give birth to their first pup, they are at least three years old and most are four years old. Most return to within a few hundred meters of where they were born and will continue to come back to that section of beach year after year.[7] Once the returning females have humped their way out of the water, they tend to move toward other females that have already arrived rather than seeking isolated places in which to give birth. The result is that, even on a long

TABLE 3.1.
Characteristics of southern and northern elephant seals (*Mirounga leonina* and *M. angustirostris*)

	Southern elephant seals			Northern elephant seals			
	Males	*Females*	*Combined*	*Males*	*Females*	*Combined*	*Sources*
Body mass (kg)							
At birth	46	40	41–43			38	1–5
At weaning	130	123	131			119–200	2, 4, 6, 7
Of breeding adults	1,500–3,510	400–600		1,704–2,275	488–700		1, 2, 4, 5, 7–11
Maximum	3,700–4,000	900		2,300–2,700	710		2, 7, 10, 11
Size ratio of breeding adults (male/female)			7–8			3–4	2, 5, 7, 9
Body length (m)							
Of breeding adults	4.7–4.9	2.7		3.8–4.5	2.6–3.1		2, 5, 8, 11, 12
Maximum	6.2	2.8		4.2–5	2.8		2, 8, 11
Size ratio of breeding adults (male/female)			1.6–1.8			1.4–1.5	2, 5, 9
Age							
At weaning (d.)			18–23			24–28	2, 4, 5, 8, 14–16
At sexual maturity (yr.)	5	3		5	3		1, 2, 5, 8, 17–19
At first mating (yr.)	9–10	3		6–9	3		1, 2, 12, 13, 18
At first birth (yr.)	10–11	4		7–10	4		1, 2, 8, 12, 13, 17, 18
Maximum lifespan (yr.)	20	23		14	17–20		2, 5, 8, 13, 15
Breeding system							
No. adults per rookery	43–165	275–1,436		350–500	1,000–2,000		7, 10, 15, 17, 19–21
No. harems per rookery			10–77			9–15	7, 15, 18, 20, 22, 23
Maximum harem size			1,350			350	7, 23

Sex ratio of mature adults (females/males)					3–4	7, 10, 17, 19, 20
Days spent ashore for breeding	49–68	27–28	7–15	91–100	25–34	1, 7, 8, 10, 15, 22
Proportion adult body mass lost during breeding	0.29–0.34	0.34–0.36		0.36	0.35–0.36	2, 4, 6, 7, 10, 16
Maximum proportion mass lost during breeding	0.52	0.42		0.46	0.41	4, 7, 10
Proportion of pups surviving the breeding season			0.84–0.98		0.60–0.87	4
Life at sea						
Duration of post-breeding foraging trip (d.)	-	62–72		120–122	66–73	1, 8, 16, 21, 24, 25
Distance traveled on the post-breeding foraging trip (km)		1,400		up to 7,500	up to 4,900	21, 27
Duration of post-molting foraging trip (d.)		255		122	243	8, 21, 24
Distance traveled on the post-molting foraging trip (km)		2,800		up to 11,100	3,900–6,800	14, 21, 26
Foraging pattern	Mainly benthic	Mainly pelagic		Mainly benthic	Mainly pelagic	24, 25, 28

SOURCES: 1, Fedak et al. (1994); 2, Deutsch et al. (1994); 3, Boyd et al. (1994); 4, Arnblom et al. (1997); 5, Bininda-Emonds and Gittleman (2000); 6, Crocker et al. (2001); 7, Deutsch et al. (1990); 8, Le Bœuf and Laws (1994a); 9, Alexander et al. (1979); 10, Galimberti et al. (2007); 11, Bonner (1994); 12, Clinton (1994); 13, Le Bœuf and Reiter (1988); 14, Le Bœuf et al. (1994); 15, Fabiani et al. (2006); 16, McDonald and Crocker (2006); 17, Haley et al. (1994); 18, Sydeman and Nur (1994); 19, Galimberti et al. (2000a); 20, Carlini et al. (2006); 21, Pistorius et al. (2008); 22, Modig (1966); 23, Galimberti et al. (2002); 24, Le Bœuf et al. (2000); 25, McConnell et al. (1992); 26, Stewart and de Long (1994); 27, Le Bœuf (1994); 28, Lewis et al. (2006).

NOTES: Values are population means unless otherwise indicated. Ranges indicate ranges in mean values reported in different studies.

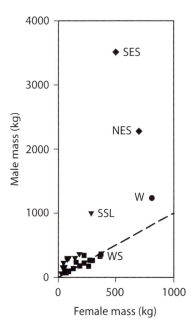

FIGURE 3.1. Body sizes (mass) of male and female pinnipeds illustrating the extreme size of male elephant seals. Each point represents the mean sizes of males and females in a single species: ◆ elephant seals, ■ other earless seals (family Phocidae), ▼ eared seals (family Otariidae), ● walrus (family Obobenidae). SES, southern elephant seal; NES, northern elephant seal; SSL, Steller's sea lion; W, walrus; WS, Weddell seal. The dashed line is the one-to-one line where male size equals female size. Data are from Bininda-Emonds and Gittleman (2000).

stretch of relatively uniform beach, the elephant seal females form clusters or groups that biologists call *harems*. Within the harem distances between neighboring females are established by vocalizations, chases, and bites so that each female establishes dominance on the small patch of beach immediately surrounding her. Larger, older females with more experience on the breeding beach are more successful at jostling for space and tend to find places in the center of the harem rather than around its edge. The harems grow in size as more and more females arrive, and at peak season tens or even hundreds of females may lie clustered together. A given harem usually occupies only one zone or section of the beach, and a breeding colony or rookery of seals typically consists of many separate harems (table 3.1).

The female elephant seals thus settle themselves on a familiar, crowded beach, each one on her own small patch of sand, surrounded by scrappy, unsociable neighbors (plate 3). They cannot forage while on land, and they will not return to the water until they have weaned their pups in about a month's time. The energy required to maintain themselves and to nurse their pups during this long fast comes entirely from fat and protein reserves accumulated during the past year's foraging at sea. They begin their fast plump and heavy, but lose about 35 percent of this mass before returning to sea. Not surprisingly, they get right down to the task before them and give birth to their pups within five or six days of coming ashore. The pups nurse for three to four weeks, gaining close to 4 kg (8.8 lb) per day, and triple their birth mass by the time they are weaned. Once the pup is born the mother does little but lie with it, either resting or nursing, until she abandons it to return to the sea. This maternal lethargy is a key component of elephant seal reproductive strategy. Since the mother cannot eat or drink during this time, any energy expended on extraneous activity is likely to jeopardize her long-term fitness by depleting her own reserves or by reducing the quality or quantity of milk transferred to her pup. The pups themselves must fast for five to ten weeks after weaning while they grow their waterproof pelage and teach themselves to swim and dive in the shallow water near the rookery. Extra fat reserves gained from rich mother's milk are essential if they are to survive through this crucial period until they can forage at sea on their own. So the mothers and their pups lie together, expending as little energy as possible unless something intrudes on their lassitude. This extreme conservation of energy explains the notable absence of frolicking elephant seal pups. Unlike the babies of most mammals, nursing elephant seal pups do not waste energy in active play or exploration.

Energy conservation and the need to return as quickly as possible to the ocean to feed also explains why elephant seal mothers wean their pups long before the pups are ready to forage for themselves. Nursing mothers usually come into *estrus* (become fertile) three to five days before weaning their pups. They mate several times during the subsequent days, and then, less than five weeks after coming ashore, they abruptly abandon their pups. Typically, the mother simply humps away across the beach, leaving what must be a very puzzled and forlorn pup squawking pitifully on the sand behind her, and she never returns to her pup again.

This cycle of returning to the rookery, giving birth within a week of arrival, nursing for three to four weeks, and then abandoning the pup to return to the sea is repeated with remarkable regularity. Females that have their first pups when only three or four years of age may skip a year before having their second pup, but most females breed every year until they die (top panel in figure 3.2). Twinning is extremely rare,[8] and so almost all of the variance in reproductive success among females comes from differences in reproductive lifespan and pup survival. Females with the lowest reproductive success are those that fail to produce any pups at all (bottom panel in figure 3.2), and by far the most common cause of this is failure to survive to reproductive age. More than 60 percent of females die before first reproduction, and the likelihood of surviving to reproduce goes down as age at first reproduction increases (figure 3.3). This is a problem common to all organisms: the longer one waits to reproduce, the higher the probability of dying before reproduction simply because disease, predation, or accident could intervene. It therefore benefits females to reproduce as early as they can, and indeed, female elephant seals become sexually mature one or two years earlier than males and begin mating as soon as they are sexually mature. Because males mature at a later age and often cannot begin mating until several years after that (more on this later), females have their first offspring four to seven years earlier than males (figures 3.2, 3.3 and table 3.1).

Given the limit of one pup per year, a female's reproductive success depends critically on her ability to successfully wean each pup. Massive pup mortality can occur if the beach is flooded by storms or high tides, and in southern elephant seal colonies, pups born on the ice or when snow still covers the beaches often starve because they sink into deep meltholes caused by their own body heat and cannot reach their mothers to nurse. Many pups die simply because they become separated from their mothers and eventually starve or are squashed beneath lumbering males. Orphaned pups are occasionally adopted by females that are not their mothers, but most females aggressively reject unrelated pups and may even kill them. Given the dangers of separation, it behooves elephant seal mothers to keep their pups close by, and this is certainly the general strategy. Mother and pup lie around all day, side by side, with the pup nursing intermittently. Should something disrupt this tranquil state, both squawk loudly until they are reunited and can resume their sleepy lethargy.

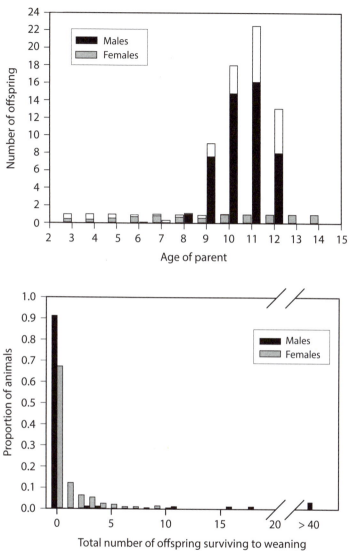

FIGURE 3.2. Reproductive success of male and female northern elephant seals on Año Nuevo Island, California. The top panel shows the mean number of offspring produced by parents of each age. For females the open bars indicate the number of pups born, and the gray bars indicate the number of pups successfully weaned. For males the open bars signify the number of females inseminated, and the black bars indicate the number of sired pups surviving to weaning. The bottom panel is a frequency distribution of lifetime reproductive success for males and females measured as total number of offspring known to have been successfully weaned over the animal's lifetime. Data are from Le Boeuf and Reiter (1988) for males born in 1964–67 and females born in 1974–75.

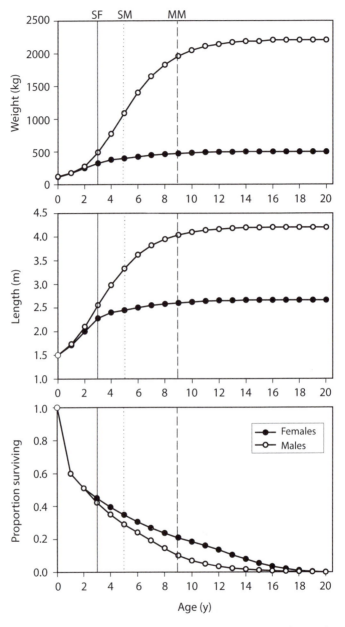

FIGURE 3.3. Estimates of survival, body length, and body mass (weight) for each age class of southern elephant seals from South Georgia. Survival is the proportion of the population surviving to that age. The vertical lines show the mean age at sexual maturity for females (SF) and males (SM) and the mean age at first mating for males (MM). Data are from Boyd et al. (1994).

Life for an adult female in the rookery thus consists mainly of lying on the beach within a very prescribed area, nursing her pup, and conserving as much energy as possible. However, as plates 1 and 3 illustrate, the rookery is a crowded place, and females are constantly bickering at each other as they jostle to maintain their small personal spaces. Larger females often displace smaller females from the best areas near the center of the harem. Relegation to the outer fringes of the colony exposes females to harassment by eager males able to escape the notice of the harem's dominant male. It also exposes females to more female–female interactions as later-arriving females join the rookery and attempt to establish their own personal spaces on the beach. All of these social interactions take energy and risk mother–pup separation. Young, inexperienced females suffer a number of disadvantages in this social milieu. They are dominated and displaced by larger, older females, and they also react more vigorously to their pup's distress calls and thus spend more energy interacting with their pups. Young mothers must also allocate some of their energy reserves for their own growth because, as in most mammals, they begin reproducing before reaching their maximum adult size (figure 3.3). The end result is that younger females are likely to find themselves in less desirable areas of the rookery where they spend more energy in social interactions and are at greater risk of separation from their pups, and at the same time they have less energy to allocate to milk production than older, larger females.[9] Not surprisingly, these young mothers are less successful at rearing their pups to weaning (top panel in figure 3.3).[10] Young mothers also deplete their own energy reserves more than larger, older females and leave the rookery in poorer condition. Because of this they are less likely to survive after breeding[11] and, if they do survive, they are less likely to give birth the following year. Clearly, females that have their first pups when they themselves are young suffer a number of disadvantages. These disadvantages or "costs" of early reproduction set the lower boundary for age at first reproduction in females. Most females first mate when they are three years old and first give birth at four years of age because, on average, these are the youngest ages at which the costs of early reproduction are offset by the countervailing benefits of increased survival to first reproduction and the potential of more breeding episodes over the female's lifespan.

After leaving the breeding rookery the females swim hundreds or even thousands of miles to areas of open ocean where they hunt for prey in deep waters. After two or three months of feeding they return to the beaches for their annual molt, which requires another month of lying onshore, fasting, and living off stored reserves of fat. To sustain this fast, females rely on the energy reserves gained during their postbreeding foraging trip. Although they mated before leaving the breeding rookery, the resulting embryos only become implanted after the molting fast so that energy is not diverted to reproduction until the females can resume foraging in the open ocean. This second foraging trip lasts eight months and ends only when the females return to the breeding beaches to give birth and begin the cycle again.

During this entire cycle of feeding at sea and molting and feeding at sea again, the females live apart from the males. Both sexes are fully aquatic while at sea and spend more than 90 percent of their time underwater, diving to depths of more than 1,000 meters (0.62 miles) in pursuit of prey. However, the females forage in different areas of the ocean and use slightly different feeding strategies than males.[12] They tend to search for food in deep areas beyond the coastal shelves, and they prey primarily on *pelagic* (open water) fishes, squid, and octopuses found at depths exceeding 400 meters (0.25 miles). In contrast, males tend to feed in shallower waters at the edge of the coastal shelf, and their dive patterns suggest that they often dive to the bottom and capture *benthic* (bottom-dwelling) animals, particularly hagfishes, rays, and skates. Although both sexes return to the beach to molt, males remain at sea longer than females on their postbreeding foraging trip and haul out to molt after almost all of the females have finished molting and have returned to the open ocean to forage.

The lives of males and females overlap substantively only in the breeding rookeries. Both come to the rookery to mate, and virtually all social interactions between them revolve around this objective. However mating is a secondary objective for females, at least in a temporal sense. Before concerning themselves with conceiving a new pup, females must give birth to and nurse their current pups. For more than 80 percent of their time ashore, they are not fertile and are unreceptive to males. Nevertheless, they are pestered several times a day by eager males that attempt to copulate in spite of the females' cries of protest and attempts to move away (plate 4).[13] Conception of the next year's pup occurs only after the female comes into estrus near the end of lactation, and during those few days she will mate several times, usually with the dominant male in the harem. As they leave the rookery to return to deep water, females may also mate with peripheral males that forcibly intercept them, but such copulations appear to be unavoidable costs of passage through a gauntlet of waiting males.

For males, life on the rookery is very different.[14] They waste no time or energy on paternal care of offspring. The pups born to females in the harem are products of the previous year's copulations and are unlikely to be the offspring of males bully enough to gain copulations this year. In terms of their own Darwinian fitness the males would gain nothing by caring for these pups, and so they ignore them altogether. What the males do focus on is gaining access to copulations with as many females as possible while at the same time preventing other males from also mating with those females. The tactics that they employ to achieve these ends are largely dictated by the distribution of fertile females. Because female elephant seals cluster close together in the same places each year and remain relatively immobile within their harems for weeks prior to coming into estrus, males can easily locate and guard large numbers of potential mates. Making things even more convenient, new females continue to join the harem as the breeding season progresses, providing males with a constantly renewing pool of mates. The key to success in such a system is clearly to become the dominant bull within a harem and keep other males away. This is indeed what male elephant seals attempt to do.

The tactic of monopolizing a harem, although simple in concept, is devilishly difficult to execute. The road to success for male elephant seals is long and arduous, and few males achieve even one successful mating in their lifetimes. In their long study of northern elephant seals in the rookery at Año Nuevo State Reserve in California, biologists Burney Le Boeuf and Joanne Reiter found that eighty-three of ninety-one males tagged as pups and followed until death failed to mate at all over their lifetimes (figure 3.2). Only nineteen of them even reached the age of sexual maturity, so, as in females, survival to reproductive maturity is a key component of fitness. Males do this less well than females (21 percent versus 35 percent at Año Nuevo, and see figure 3.3 for southern elephant seals on South Georgia Island) mainly because they mature about two years later than females. Females also mate as soon as they are sexually mature and almost always give birth to a pup the first season that they appear in the breeding rookery, whereas males seldom achieve a mating until they have been on the breeding beaches for several successive years. Males therefore have to survive to much older ages than females if they hope to father offspring (table 3.1). Even among the few males that survive long enough to achieve a mating, reproductive success is highly skewed (figure 3.2).[15] The three most successful males in Le Boeuf and Reiter's study fathered ninety-three, eighty-two, and forty-one offspring over their lifetimes and were successful in seven, three, and four seasons respectively. Considering that none of the others fathered more than twenty pups and 91 percent of males never mated at all, the inequity among males is extraordinary.

The stakes are clearly high for male elephant seals. Chances are that they will have no reproductive success at all, but if they are successful, they can be spectacularly so. The key is to become the dominant male within a large harem of females, a social status that biologists call the harem master or alpha male (plate 1). Harem masters monopolize the great majority of copulations and can achieve spectacularly high reproductive success. It is not unusual for a dominant male to control a harem of fifty to one hundred females and to successfully fertilize 80 percent to 100 percent of females within that harem. In even larger harems secondary and peripheral males share relatively more of the copulatory success, but the dominant or alpha male is still likely to fertilize more than three-quarters of the females.[16] A study of southern elephant seals on Sea Lion Island in the Falkland Islands provides a good example of the benefits of holding a harem.[17] Researchers observed copulations and used microsatellite genetic markers to determine the paternity of pups in this colony over two breeding seasons. They discovered that harem masters achieved 94 percent of the copulations and fathered 90 percent of the pups born. The number of pups fathered by harem masters ranged from 18 to 125, whereas males that were not harem masters fathered between 0 and 6 offspring. Of the males present in the rookery, 72 percent were never observed mating and fathered no offspring. Since some bulls are able to hold harems for several breeding seasons (a maximum of six years in southern elephant seals), the distribution of paternities would likely be even more skewed if translated into lifetime reproductive success as in Le Boeuf and Reiter's study of northern elephant seals.

Harem masters are usually larger, older bulls that have many years of experience interacting with other males on the breeding beaches. Males begin arriving on the breeding beaches before the first females haul out. They proceed to threaten each other by rearing up on their front flippers, swinging their heads back and blowing air through their long noses to produce a series of deep, burbling honks (longer noses make deeper and more impressive sounds).[18] These displays very effectively show off their size and power. Males challenge each other one-on-one, and the smaller male will usually back down, especially if he is a late arrival and the other male has already claimed that section of beach. However, displays do escalate into outright fights in which the males push each other and swing their heads trying to jab each other with their large canine teeth. Their thick neck hide absorbs these blows, but most contending males are bloody and scarred by the end of the season. By the time the majority of females begin arriving, the males have established a dominance hierarchy among themselves, and the dominant males take control of the groups of females as they coalesce. Southern elephant seal males may actually herd females to keep them in the harem, but northern elephant seal males seem to accept the female distribution as it comes. By and large, for both species, keeping control of a harem means intimidating other males more than bullying females. Females benefit from this behavior because it reduces the amount of harassment they receive from other males, and this is one reason why they prefer sites in the center of the harem close to the dominant male.

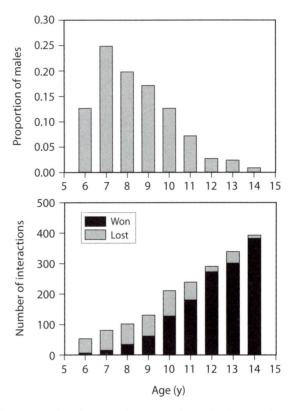

FIGURE 3.4. The age distribution of male southern elephant seals on the breeding rookery (top panel) and the mean number of aggressive interactions won and lost by males of each age (bottom panel). Data are from Sanvito et al. (2008) and are averaged over eight breeding seasons at Sea Lion Island in the Falkland Islands.

The success of males in the behavioral interactions that lead to dominance increases with age (figure 3.4). Older males initiate these interactions more often and win them more often. They also settle more of the disputes with vocalizations alone, thus avoiding costly fights and chases. The main reason why older males are more successful is because they are larger.[19] Size conveys strength and inertia that contribute to success in physical contests with other males, and larger males also have an energetic advantage because of their lower mass-specific metabolic rates (i.e., it takes less energy per unit mass to power a large animal than a small animal). The latter advantage allows larger males to accumulate more energy reserves while foraging at sea and to sustain a longer fast on the breeding beaches. They can also afford to devote relatively more energy to mating and to dominance interactions than their smaller rivals. Because of the advantages of large size, male elephant seals continue to grow for many years after puberty (figure 3.3), and sexually mature males that have not yet reached their full adult size (often called subadults) seldom even compete in the dominance interactions. They may hang around the edges of harems and try to sneak copulations when the harem master is distracted or when females leave the harem, but they are seldom successful (plate 5). Not only are they physically intimidated by the larger, older males, they also have the disadvantage of having to devote energy to growth. This means that they come to the breeding beaches with a lower proportion of body fat than older males and so cannot sustain as long a fast, and those that attempt to breed pay a price in reduced subsequent survival and reproductive success.[20]

So, the key to being a successful male elephant seal is to live long enough and grow large enough to become a dominant harem master. If not a harem master, a male may gain at least some reproductive success by rising high enough in the dominance hierarchy to become a secondary male within a large harem in which the dominant male is not able to defend all the females. A third tactic is to achieve dominance within the horde of peripheral males that attempt to intercept females departing the harem. However, because the females have already mated within the harem, these matings are unlikely to lead to paternity for the male. Dominance on the periphery is therefore not the best tactic for males in the long term, and most males in this category are younger and smaller than the harem males. Because dominance rank is the best predictor of reproductive success and body size is the strongest determinant of dominance, it behooves males to grow fast and to keep growing until they are large enough to compete for mates. This is exactly what they do. Male southern elephant seals are larger than females at birth (table 3.1), and both southern and northern males grow much faster than females, especially after age two (figure 3.3).[21] They also keep growing long after female growth has reached its plateau. Southern elephant seal males achieve larger adult sizes than northern males mainly because they continue growing until age twelve, whereas growth plateaus at age eight or nine in northern males. While younger females are putting their energy into producing their annual pups, younger males are putting theirs primarily into continued growth.[22] They visit the breeding rookery for increasingly longer periods as they get older and larger and gradually begin to participate in the one-on-one male dominance interactions. As their growth slows, they are finally able to accumulate sufficient reserves of fat to sustain high levels of aggressive and sexual activity over the long breeding season. These older males arrive at the rookery early and may stay until all the females have left. Their massive reproductive advantage over other males is a consequence not only of their ability to attain dominance but also of their ability to retain their position throughout the long breeding season as females come and go. Very few males are able to survive long enough and grow large enough to attain this exalted status, but for those that do, the rewards in the currency of Darwinian fitness are huge.

How different this male perspective on life is from that of the females lying on the beach with their pups. Males striving to physically intimidate other males and to inseminate as many females as possible are bound to come into conflict with females whose main goal is to conserve energy while they give birth and nurse their pups. Indeed, the rookery is rife with sexual conflict. When females first arrive at the rookery they are heavily pregnant and certainly not fertile. Nevertheless, peripheral males often attempt to mount them as the females hump their way from the water to the relative safety of a harem. The gauntlet is even more daunting for females leaving the rookery and attempting to return to the sea at the end of their breeding haul-out. By this time they are weak and hungry, having lost about 35 percent of their body mass through their long fast, and they have no need for further matings because they have already mated several times within their harems. Nevertheless, eager males converge on the departing female, squabbling among themselves to establish first access, but paying no attention to the female's obvious protests. At Año Nuevo rookery, researchers Sarah Mesnick and Burney Le Boeuf[23] observed that almost all departing females were mounted and most were forced to copulate before they reached deep water. Some endured as many as five mountings as they crossed the sand and intertidal zone. These mating attempts were often extremely aggressive. The females experienced head slams and neck bites so severe that they sometimes sustained broken ribs, head injuries, organ damage, and internal hemorrhaging. Of the seventeen female deaths that Mesnick and Le Boeuf observed on the breeding rookery, eleven were caused by traumatic injuries inflicted by males during mating attempts as the females attempted to reach the sea. Most females were not killed during these mating attempts (Mesnick and Le Boeuf estimated the overall mortality rate to be about 1 per 1,000 over twenty years of observations), but the frequency of debilitating injuries that were not immediately lethal was probably considerably higher than this.

Females experience less extreme sexual harassment while they are within harems, but they must nevertheless endure repeated mating attempts by the harem master. Researcher Filippo Galimberti and colleagues observed interactions between male and female southern elephant seals at Sea Lion Island for several breeding seasons and reported that the males were so eager to mate that they even attempted to mate with dead females, small juveniles, and other seal species.[24] They constantly showed aggressive behaviors such as pushing, chasing, and biting when interacting with females within their harems and aggressively herded their females by putting their chests on the females' backs or slamming the females with their chests. Females within harems endured such aggressive interactions two or three times each day, and about three-quarters of these interactions occurred before the females came into estrus. Northern male harem masters seldom attempt to herd their females, but they do make repeated attempts to mate. In both species these mating attempts are far from gentle wooing. It seems that the males cannot assess a female's reproductive condition (i.e., whether or not she is fertile) by visual or olfactory cues that can be detected from afar. Instead, males assess a female's receptivity the direct way—by humping up to her, rearing up, pushing her down, biting her neck and trying to copulate (plate 4). Females almost always protest this treatment by vocalizing loudly, moving away, shaking their hind quarters, and even biting. However, they do so less vigorously when they are in estrus, and this apparently is the clue that the harem masters are looking for. (Northern females are a little less subtle than southern females and may actually adopt a receptive posture called *lordosis* when they are in estrus.) Harem masters are generally deterred if their females protest vigorously enough and will dismount and move away without achieving intromission. However, forced copulations do occur. Females near the periphery of the harem suffer additional harassment from subordinate and peripheral males. In these cases their loud protests may attract the attention of the harem master, who then rushes over and attempts to disrupt the mating, often at even more risk to the female and her pup.[25]

Clearly, life in the elephant seal rookery is not a romantic paradise for either gender. Sexual interactions are aggressive and often brutal. Most males cannot get enough sex, and females cannot help getting more than they want or need. Both males and females spend valuable energy on copulation attempts that cannot possibly lead to fertilizations. For females these superfluous sexual interactions carry the risk of injury to herself or her pup. Mothers and pups also become separated in the melee of male eagerness and female protests, and if they fail to find each other again, the pup faces almost certain death. Everyone risks injury when the males chase and battle with each other to monopolize mating opportunities, whether within a harem or among the hordes of peripheral males attempting to intercept females on the way to and from the water, and the huge size difference between males and females increases the risk to both females and pups.

Given the aggressiveness and heedlessness of the galumphing males, it is surprising that the mortality rate in the breeding rookeries is not higher than it is. In fact the majority of pups and almost all the adults survive until the end of the breeding season (table 3.1). The saving grace is that both sexes are fasting and have to be miserly with energy, so they spend most of their time in the rookery lying around and resting. Male southern elephant seals spend about 85 percent of their time resting, whereas females are at rest about 89 percent of the time. Estimates for northern elephant seal rookeries are similar or even a little higher: males rest 85 percent to 95 percent of the time, whereas females rest 92 percent to 95 percent of the time.[26] At the height of the breeding season when most of the females have come ashore and given birth and when the dominance relationships have been well established within the social hierarchies of both sexes, the rookery presents a picture of lethargy much of the time, with rank upon rank of prone bodies lying motionless on the sand. This fragile peace, enforced by the need for both sexes to conserve energy, allows the majority of pups to nurse, grow, and survive in spite of the heedless aggressiveness of the huge adults around them. The aggressive interactions of the adults appear brutal, but they generally cause only minor injuries, particularly bite wounds, and are seldom lethal. Most adult mortality occurs at sea when, among other hazards, seals fall prey to sharks and killer whales.

And so we come full circle in the lives of male and female elephant seals. To be sure, they have much in common. Both forage at sea and come to land twice a year, once to molt and once to breed. They share the same breeding and molting beaches, and for the former purpose, occupy the beaches at the same time. However, within these broad general parameters, the two sexes have evolved very different life histories. They grow at different rates and differ enormously in age and size at first reproduction. They feed in different areas of the ocean, using different foraging techniques and eating different prey species. Even when together in the breeding rookeries males and females have different primary objectives (females to give birth and nurse their pups and then to conceive their next pup; and males to mate with as many females as possible), and they live within separate, sex-specific social hierarchies. Females maximize the number of offspring they contribute to the next generation (our measure of their Darwinian fitness) by mating for the first time between the ages of two and four and then producing one pup each year for as many years as possible. Theirs is a steady accumulation of reproductive success limited mainly by reproductive lifespan. Males take a very different path. Their Darwinian fitness depends on success in competition with other males for access to fertile females, and success in this competition depends primarily on body size. They are likely to have to forgo mating for many years after puberty, and more than 90 percent will leave no offspring at all. Few live long enough or grow large enough to become harem masters, but those that do are likely to be spectacularly successful, leaving up to ten times more offspring than the most successful female. This mating system is the most extreme example of *polygyny* known in vertebrate animals[27] and, to my knowledge, in any animal species. It is a push-shove, boom-bust world of male competition in which large size pays such a premium that male body size in elephant seals has increased well beyond that of any other mammal in their order (Pinnipeds)—and well beyond the maximum sizes of their own females.[28]

For their part, female elephant seals must coexist with huge, belligerent males that have great potential to injure them or their pups. As far as we can tell, the males provide nothing for the females other than sperm (and the genes their sperm contain) and some degree of protection from harassment by other males. In addition, because of their large size and rapid growth, males require more than their share of resources[29] and so could be major ecological competitors of females if the sexes shared the same foraging areas.[30] Females live within a dominance hierarchy of their own, and, as in males, larger and older females gain some reproductive advantages. However, the effect of body size on reproductive success is very much weaker in females than it is in males, and the advantages of large size are offset both by the benefits of reproduction at an early age and by diversion of energy to pregnancy and lactation rather than growth. Thus, in spite of the dangers and costs of coexisting with large, aggressive males, females stop growing earlier than males and reach much smaller maximum sizes. The most obvious manifestation of these differences in male and female reproductive strategies is the most extreme sexual size dimorphism among mammals and one of the most extreme examples of male-larger sexual size dimorphisms in the animal kingdom. This huge disparity in body size between males and females is associated with equally impressive differences in the age schedule of reproduction, in the distances and directions of migration to ocean foraging areas, in both the dive patterns used for foraging and the types of prey captured, and in the types and frequencies of social interactions in the breeding rookeries. Male and female elephant seals are truly different throughout their adult lives.

Similar harem-based, polygynous mating systems can be found in many large mammal species although not to the extreme seen in elephant seals. Notable examples include the various species of sea lions and fur seals, which are in a separate seal family (Otaridae) and represent a separate evolutionary origination of this type of mating.[31] Many ungulate species (deer, sheep, bison, and antelope) have also evolved harem-based breeding systems, as have several primate species, and the parallels between these species and elephant seals are often striking.[32] The key to these systems is that males are able to monopolize mating opportunities with females because fertile females are found in clusters or aggregations that can be defended by large, belligerent males; and the signature traits are marked sexual dimorphism in body weight, with males being the larger sex, and male-limited weaponry such as horns, antlers, tusks, and enlarged canine teeth. Intense selection on males to outcompete their rivals (sexual selection) is certainly the main reason why males become so much larger than females,[33] but we must be careful not to ignore the importance of selection on females as well. Females typically live within a dominance hierarchy of their own and gain advantages from increased size, but as in elephant seals, the benefits of large size for females are very modest when compared to the potential gains that large males can achieve. This is mainly because females give birth to only one or two offspring per breeding attempt and maximize their lifetime offspring production by producing one high quality offspring in each of many consecutive breeding seasons rather than by increasing the number of births per season. The net effect of this strategy is that females begin breeding at an earlier age and smaller size than males, and they divert energy into offspring production rather than growth, so that they remain much smaller than males throughout their adult lives. Elephant seals serve as the most extreme example of this pattern of sexual differences, but they are unique only in the magnitude of their sexual dimorphism and the scale of their ecological and geographic segregation. Similar patterns are repeated to a lesser degree in many large mammal species, and as we shall see in subsequent chapters, traces can be discerned in the patterns of sexual differences found in polygynous species in other taxa as well.

CHAPTER 4

Great Bustards

Gorgeous Males and Choosy Females

Great bustards (known to science as *Otis tarda*) are iconic birds of open, rolling grasslands and agricultural areas across a wide swath of temperate Eurasia.[1] They are huge birds with features reminiscent of both cranes and grouse, and they are most often seen foraging quietly for insects, seeds, and herbaceous vegetation in agricultural fields and fallow areas.[2] Although they are probably best known for the gorgeous mating displays of their males, the "great" in their name refers not to the impressiveness of those displays but, rather, to the huge size of the adult males (table 4.1). Historical records credit males with weights of up to 24 kg (53 lb), and although weights larger than 15 kg (33 lb) are now uncommon, male great bustards still qualify as the heaviest living flying birds, a distinction they share with African kori bustards.[3] Not so for females, however. Female great bustards are much smaller than males, averaging only 80 percent as tall and one-third the weight of their mates.[4] This huge discrepancy in size makes great bustards the unchallenged bird champions in terms of sexual size dimorphism. As in elephant seals, this extraordinary sexual dimorphism is associated with a highly polygynous mating system in which larger males are successful in mating competition with other males, while small males are excluded from mating at all. However, the nature of the mating competition differs considerably between these species. Rather than defending harems of females by physical intimidation, great bustard males congregate in traditional mating areas and attempt to attract females by

TABLE 4.1.
Characteristics of great bustards, *Otis tarda*

	Males	Females	Sources
Body mass (kg)			
Mean adult	8.9–12.0	3.8–4.4	1, 2, 3
Maximum	13.0–24.0	5.0–5.2	1, 2
Skeletal size (cm)			
Mean wing arch[a]	61.7–62.8	48.6–49.1	1, 2, 3
Mean tarsus[b]	15.3–15.8	12.0–12.5	1, 2, 3
Range in height	90–105	75–85	4
Range in wingspan	210–250	170–190	4
Age			
Mean at independence (mo.)	9.5	10.8	5
At first appearance on the lek (yr.)	2–4	1–3	5, 6, 7
At first breeding (yr.)[c]	4–6	1–4	5, 7, 8, 9, 10
Maximum lifespan (yr.)[d]	15, 20	20	5, 11
Migration and spatial distribution			
Mean distance between natal site and breeding lek (km)	14.1–12.5	4.3–4.0	5, 11

SOURCES: 1, Alonso, Magaña et al. (2009); 2, Johnsgard (1991); 3, Lislevand et al. (2007), Székely et al. (2007); 4, The UK Great Bustard reintroduction project, http://www.greatbustard.com/identification.html, accessed March 23, 2010; 5, Alonso et al. (1998); 6, Alonso and Alonso (1992); 7, Morales et al. (2002); 8, Ena et al. (1987); 9, Alonso et al. (2004); 10, Morales et al. (2000); 11, Morales et al. (2003).

[a]A standard measure of wing length in birds. It is actually the length of the distal or outer portion of the wing, beyond the carpal or "wrist" joint, measured along the dorsal side of the wing from the carpal joint to the end of the longest primary feather at the wing tip.

[b]A standard measure of leg length in birds, based on the length of the long bone of the lower leg, the tarsus.

[c]First breeding is first copulation for males and first egg laying for females.

[d]Estimated for wild populations. Maximum lifespan of males in captivity is at least thirty years (Johnsgard, 1991; Morales et al., 2003).

performing elaborate displays. The traditional mating areas contain no extra or unusual resources for females and seem to serve no purpose other than providing a traditional site for males to display. The females visit the display sites only to choose a mate, and the close proximity of the males enables them to comparison shop. (The entire process is rather like speed dating for birds.) This type of mating system, with its enhanced opportunity for female choice, is called *lekking*, and the aggregate of displaying males is the *lek*. Competition among the lekking males is intense, and in most lekking species only a few males garner the great majority of matings.[5] Lekking males compete with each other indirectly by trying to attract females through their displays, but they also compete directly by trying to physically intimidate other males to gain the most favorable display sites. If the aggressive interactions between males occur on the ground (as opposed to in flight), larger males usually have the advantage. As a result, in many bird lineages, the species with the largest males and the greatest magnitude of sexual size dimorphism are those with lek mating systems and ground displays. The bustards (the family Otididae) are a good example of this trend: great bustards are the only bustard species with a true lek mating system, and they have the largest males and by far the greatest sexual size dimorphism in the family.[6]

Given this association between lekking and extreme sexual dimorphism, formation of the leks in late winter seems a good starting point for our examination of the separate lives of male and female great bustards. The annual cycle of bustard behavior varies somewhat across its large geographic range, but for the most part my descriptions are based on the populations in Spain and Portugal. These are the largest and healthiest populations and have been studied intensively for several decades, so they provide the most detailed and complete record of bustard life histories.[7] In these populations males begin to congregate in their traditional lekking areas in February. The mature males undergo a breeding molt and replace their rather dull nonbreeding plumage with bright, ivory-white throats, deep mahogany-brown plumage at the base of their necks and on their breasts, and long white moustachial feathers that hang down from the sides of their beaks like whiskers (plate 6). This distinct breeding plumage develops gradually as males get older, seldom reaching full development until they are five or six years of age. The moustachial feathers may continue to lengthen even beyond that age. Mature males also develop thicker necks and inflatable esophageal sacs that enable them to expand their necks to better display the rich mahogany of their plumage.

As their breeding plumage becomes more obvious, the males behave with increasing aggression toward other males in the flock. They threaten, chase, and even physically attack one another (plate 7). Typically one male will approach another and display his breeding plumage by raising his moustachial feathers beside his bill, partially inflating his esophageal sac, and raising his tail over his back to reveal the brilliant white feathers underneath. At the same time, he lowers his folded wings and twists them outward to display his white wing feathers. These threat displays generally last less than a minute, but they are sufficient to cause most rivals to back down and flee. If the challenged male displays back, the two males often strut back and forth, literally sizing each other up. If neither male concedes, the contest is decided by physical combat. Typically, the two males peck at each other's faces and eyes until one male grabs the other's bill. So joined, they push and shove each other, sometimes for over an hour. When the contest finally ends, the males may be so exhausted that they cannot fly and one or both may have suffered injuries to their bills and loss of or damage to their precious moustachial feathers.

These one-on-one physical contests serve to establish a dominance hierarchy among the males in which ultimate rank depends on age, body size, and physical condition.[8] Younger males form all-male flocks near the lek but do not participate in the dominance displays. Among the fully adult males, those that are smaller or in poor condition may fail to produce full breeding plumage even though they may be old enough to do so, and these males are not successful in the dominance contests. Thus, although large flocks of males initially congregate in the lekking areas, only the larger, healthier, more mature males engage in the serious business of establishing dominance in the breeding hierarchy. Contests between these males are generally won by the contestant that is heavier and in better condition.

While the males are busy sorting out their relationships with each other, the females begin to appear on the lekking area, and this greatly complicates the social situation. The females arrive singly or in small flocks with the intent of selecting mates. The purpose of the lek is now revealed. The males turn their attention to the visiting females and try to entice them away from other males. The nature of the male-female encounters is varied and fluid, especially early in the breeding season when males have not yet completely sorted out their dominance relationships. Sometimes a group of females encounters a group of males, and the two sexes travel together creating temporary mixed-sex breeding flocks. If one male succeeds in enticing the females away from his confreres by judicious herding and display, a temporary harem-like grouping can occur with one breeding male and several females. As the season progresses displaying males tend to space themselves out in the lekking area, moving from display site to display site, with no fixed territorial boundaries. Lone females may approach a displaying male, allowing for isolated, one-on-one mating interactions, or the male may attract a group of females, which then follow him as he displays, again forming ephemeral, harem-like groupings.

For all the variety of male-female groupings on the lek, the one consistent feature is the necessity of male display as a prerequisite for successful mating. To entice a female to mate with him, a male has to "strut his stuff" vigorously and often. The male courtship display is an exaggerated version of the display males use when they interact with each other (plate 8). The prized moustachial feathers are raised to fully vertical position, the brilliant white wing and tail feathers are fully exposed by contortions of wings and tail, and the esophageal pouch is fully inflated to reveal the full glory of the chestnut plumage and deep blue-gray skin patches slicing down the sides of the neck. The displaying males resemble colorful, walking balloons, and indeed, biologists call this the balloon display. The males strut stiffly, circling around their potential mates. Around and around they go as the female walks slowly, inspecting their every move. From time to time the male may place a wing over the female's back, perhaps to persuade her to crouch in mating position. At other times he turns away from her, presenting his brilliant white posterior. If she pecks at this, the chances are good that she will eventually accede to a mating.

During the height of the breeding season males display repeatedly from dawn to dusk, resting only during the heat of the day or when the temperature drops below about 12°C. They display to attract females from a distance and, once the females have approached, to persuade them to mate. Both the number of females a male attracts and the number that actually mate with him are strongly correlated with the amount of time he spends displaying. Females pay attention to the quality of the male's display, particularly his neck plumage and moustachial feathers, but these would be all for naught if he could not sustain the display for long enough.[9] Males whose breeding plumage is not fully developed seldom perform the full balloon display, and if they attempt to attract a female, they are likely to be interrupted by more dominant males. The utility of the male dominance hierarchy now becomes obvious. Dominant males interrupt courtship and mating attempts by subordinate males but are able to sustain their own courtships without interruption. Just as we saw in elephant seals, high dominance rank in the male hierarchy translates into success in male-female interactions.

The advantages of size and condition so evident in competition for dominance extend through the period of mate attraction. Once the peak of the breeding season arrives, males on the lek spend almost all of their time displaying and very little time feeding, and by the end of the breeding season some are so weakened from starvation and exhaustion that they cannot fly.[10] To sustain this mating effort, they have to gain prodigious amounts of weight. Males that are ultimately successful on the lek typically weigh 20 percent to 30 percent more by the peak of the breeding season than they did in late winter.[11] The males that are larger at the onset of the breeding season are able to gain the most weight, probably because their large size enables them to establish dominance in less time and with a lower expenditure of energy than would be required of smaller, lighter males. The result is a positive feedback between large size and condition that exaggerates the size differences present among males at the start of the breeding season and increases the intensity of selection for large size in males.

Clearly female great bustards demand a lot from their mates. The congregation of males on breeding leks enables females to comparison shop, and only the largest, healthiest, and most vigorous males are able to sustain displays that pass muster. These supermales are able to mate with many different females. Observers have recorded up to five mates for a single male in a given season and this is surely an underestimate of the maximum mating success. On the other side of the coin, most males lose out completely. Less than 15 percent survive to reproductive age, and of those that do, only about one third succeed in copulating in any given year.[12] This variance among males probably compounds over the lifetimes of individual males because the successful, dominant males tend to retain their positions for several years. Only males that are able to sustain high growth rates as juveniles and maintain high body condition as adults have a chance in the vigorous competition for mates. Even the largest, heaviest males must also have the behavioral repertoire and social skills to both intimidate rival males and entice discriminating females. In the currency of Darwinian fitness the rewards are potentially very large for the few males that combine these attributes, but for the majority of males, failure is absolute.

In contrast to all the stress and drama of life on the lek for males, female great bustards take a very measured and calm approach to mating.[13] Most spend only a few days on the lek, during which time they walk slowly among the males, observing displays and engaging from time to time in abortive courtships. They endure none of the frenzied and potentially injurious physical harassment that greets female elephant seals that venture onto the breeding beaches. In the great bustard world females choose their mating partners, and males compete among themselves to be the chosen ones. When dealing with females on the lek, persuasion, not intimidation, is the key. Typically, a female will approach a displaying male from afar. In response, the male usually increases the intensity of his display. If the female is impressed, she will continue her approach until within pecking distance of the male. She may continue to walk slowly while the male keeps pace, circling her and performing his balloon display. Most often, this is as far as it goes. The female eventually walks away without consummating the relationship. However, each female inevitably finds a male to her liking and, after putting him through his paces, crouches to permit the brief copulation. Once mated, she loses interest in the displaying males and walks away to find a suitable spot to lay and brood her eggs.

The female's short-lived visit to the lekking area is likely to be free of strife unless she finds herself competing with other females for the attention of a given male. If more than one female becomes interested in a displaying male, the most dominant female may attempt to displace the others from the male's attentions. These are very mild interactions when compared to the drag-out fights between males. The dominant female may succeed in being the first to mate with the male of interest, but the other females need only wait until she has mated to have their chance. Unlike males, females do not prevent other females from eventually mating, and it is likely that every female that seeks a mate is successful in finding one.

For females, the major impediment to reproductive success is failure in chick rearing, not failure to find a mate. Choice of a location for her nest is one of the most important decisions that a female great bustard makes in her lifetime.[14] The growth and survival of her chicks, and hence her own Darwinian fitness, depend strongly on this choice. Prior to the advent of mechanized agriculture and the associated anthropogenic threats to bustard eggs and chicks, the primary cause of nest failure was predation. Even today many eggs and chicks are lost to predators such as dogs, foxes, ravens, and crows, and the mothers themselves are vulnerable to predators while brooding their eggs. To minimize these risks female great bustards prefer nest sites that conceal the nest from predators while at the same time allow the brooding female to survey the surrounding area for danger. If the vegetation is too tall or dense, the nest can be ambushed. Even if the predator is detected, dense vegetation may impede escape. Alternatively, nesting in open sites such as plowed fields leaves the nest too visible. In the Spanish populations agricultural fields planted in cereal crops or that are fallow seem to provide the best compromise between visibility and concealment and are preferred as nest sites. Successful females also select nest sites that offer some protection from inclement weather. Newly hatched chicks have poor thermoregulatory capabilities, and cool, rainy weather during the hatching period can cause high chick mortality. In central and northwestern Spain, sites on southeast-facing slopes that catch the morning sun but are sheltered from the cool, prevailing northwesterly winds are preferred for this reason.

Once a female has selected her nest site, she scrapes a rough depression in the grass or bare ground and lays one to three eggs (very occasionally four) that she then incubates for three to four weeks.[15] When they hatch, the chicks are *precocial* balls of mottled down. They are soon following their mother through the vegetation as she pecks in the grass for insects and other tasty food items, including the occasional frog or rodent (plate 9).[16] Initially the mothers provide all of the food for their chicks by feeding them bill to bill, and the chicks continue to receive extra food from their mothers at least until they are nine or ten months old.[17] To accommodate the food requirements of their chicks, great bustard mothers must have very large home ranges (averaging 3 km [1.9 miles] in diameter) that encompass prime foraging areas.[18] This need for rich foraging areas away from competition with other bustards explains why females often move many kilometers away from the breeding leks to brood and rear their chicks.[19] By late summer or early autumn the isolated families of females and chicks begin to coalesce with females without young to form mixed over-wintering flocks. By this time insects are becoming scarce, and the birds rely increasingly on herbaceous vegetation and seeds as the winter progresses.[20] As spring approaches the adult females in the flock become increasingly aggressive toward the now large chicks, and the chick-mother bonds gradually weaken. Most male chicks are independent of their mothers by ten months of age and females about a month later (table 4.1), although mother-chick associations occasionally persist for as long as seventeen months.[17]

Female great bustards adopt a number of behaviors to protect their nests and chicks during their long period of maternal care. Brooding females usually crouch on the eggs and remain immobile when danger approaches, relying on camouflage. If the threat is a predator and it approaches too closely, the female may attempt to lure it away by using a prominent distraction display in which she walks away from her nest or chicks with her wings lowered and tail raised. Occasionally mothers may even defend their nests and chicks aggressively by threatening or even attacking potential predators, including humans and other bustards (plate 10). These protective and defensive tactics work well against natural predators, but unfortunately they are of little avail against the modern agricultural machinery that inadvertently destroys many nests and chicks each year.

In spite of all of their efforts to feed and protect their young, female great bustards have little success in rearing their chicks. A study by researchers Vincent Ena, Ana Martinez, and David Thomas provides striking documentation of the difficulties faced by great bustard mothers.[21] The researchers followed the fates of nests and chicks in a comparatively large and healthy population of great bustards in northwestern Spain. At least 92 percent of the females in this population produced a clutch but only a fifth of the eggs and a third of the mothers produced chicks that survived to the end of August. Further, although most of the females laid two or three eggs, only 8 percent had more than one surviving chick. By far the greatest cause of reproductive failure was nest destruction by agricultural machinery, mainly during harvesting, and this accounted for total loss of half of all nests. Eggs that escaped this fate had relatively high hatching success (90 percent), but only 43 percent of the chicks that hatched managed to survive the summer. The rest were killed by predators (mainly ravens, hooded crows, kites, foxes, and dogs) or farm machinery. Because of this huge chick mortality the annual productivity for this relatively healthy population was only forty-four *fledglings* for every one hundred adult females.[22]

Chicks that escape these early dangers face the challenge of obtaining enough food. Food scarcity seems to be an ever-present threat for great bustards, and females often fail to obtain enough food to support themselves and their chicks. Researchers Manuel Morales, Juan Alonso, and Javier Alonso conducted an eleven-year study of female reproductive success in a population of great bustards in northwest Spain, and their results starkly illustrate the crucial dependence of bustards on their food supply.[23] In this region breeding females increase their body mass by about 16 percent just prior to egg laying, and they depend on the spring flush of annual herbaceous vegetation to do so.[24] When winter rainfall is above average the density of herbaceous vegetation is high, and more females attempt to breed. Females breeding in those years also lay more eggs, and the eggs have higher hatching success. Abundant winter rain also increases the density of the annual grasses, and these provide food for the grasshoppers and crickets that become the main food for great bustard chicks. Chick growth and survival during the critical first three months after hatching depend strongly on food abundance, and so abundant summer food means more and larger chicks fledging in fall. The net result is that abundant winter rains lead to more breeding females, more eggs per female, greater hatching success of eggs, and better chick growth and survival. Conversely, in years of low winter rainfall and hence low food abundance, bustard productivity falls close to zero with as few as four chicks produced per one hundred adult females.

The energetic costs of producing eggs and rearing chicks probably contribute to the extreme sexual size dimorphism in great bustards by favoring relatively smaller body sizes in females than in males. The nub of this argument is simply that females have to provide food for both themselves and their chicks, and the less they need for their own maintenance, the more is available for egg production or to be fed directly to the growing chicks. When food is scarce, as it often is for great bustards, larger individuals may be not be able to obtain sufficient calories to both sustain themselves and allocate energy to reproduction.[25] The trade-off between self-maintenance and reproduction is manifest every time a brooding mother forgoes a feeding foray to sit longer on her nest or feeds a prey item to her chick rather than eating it herself. Eating it herself increases her own body condition and enhances her chances of reproducing in the coming spring but at the cost of diminishing the potential growth, survival, and future reproductive success of her chicks. As a consequence of this trade-off, great bustard females that successfully rear a chick in one year seldom have sufficient energy reserves to breed successfully the following spring.[23]

Clearly the reproductive lives of male and female great bustards are very different. Males face a lifetime of zero reproductive success unless they can grow large enough to best their peers in contests for dominance on the lek. Having done that they are still out of luck unless they can attract females and persuade them to mate. To do all of this they need to be big, aggressive, energetic, and gorgeous. Achieving this exalted status is a long, slow, and risky process in which few males ever succeed. Males have higher mortality rates than females at all ages,[26] and many of those that do survive fail to achieve the minimum body size and condition necessary to gain dominance on the lek and sustain the extravagant balloon displays necessary to attract females.[27] This is a high-stakes game, and there are no rewards for mediocrity. It is not surprising, therefore, that achieving mating success is the focus of male behavior and life history. By comparison females have no difficulty securing matings, and they are quite picky about selecting a mate, preferring males whose elaborate mating displays reveal good health, good body condition, and strong competitive abilities. By being so selective females are complicit in the evolution of large male body size and stunning courtship displays, although their own fitness is best served by cryptic plumage and much smaller body size.[28] In sharp contrast to males females devote most of their time and energy to nurturing offspring. They begin breeding about two years earlier than males and thereafter most lay eggs every year (table 4.1).[29] Their success at rearing chicks is poor at first, but by repeated efforts most have successfully fledged a chick by the time they are four or five years old.[30] Females play the game for much lower stakes than males because their total reproductive success in any given year can be no higher than the number of eggs they can lay and is seldom higher than one successful chick. Their lifetime reproductive success depends on good parenting skills, repeated breeding attempts, and a long reproductive life—not the boom or bust of competition in a few mating seasons.

Not surprisingly these sex-specific reproductive strategies have resulted in marked differences between the sexes not only in adult plumage and body size but also in juvenile growth rates and early survival.[31] Males grow faster than females and continue to grow for a longer period (figure 4.1). Female growth slows dramatically at about seven weeks of age. By comparison, male growth rate does not show a similar decline until ten to twelve weeks of age, and by that time males weigh twice as much as females. Both sexes continue to gain weight slowly, but males gain proportionally more, almost tripling their weight by the time they are fully mature. To support this differential growth male chicks have to eat much more than female chicks. By ten days of age they require about 16 percent more energy than females, and this differential increases with age. Not

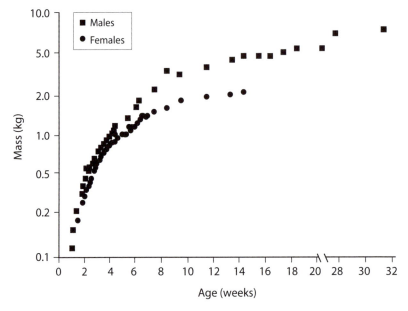

FIGURE 4.1. Growth rates of captive-raised male and female great bustards. Note that mass (weight) is shown on a logarithmic scale. Redrawn from Johnsgard (1991). Original data from Heinroth and Heinroth (1927).

surprisingly, great bustard mothers feed their male chicks much more than their female chicks. Spanish researcher Juan Alonso and his colleagues documented this difference at the Villafáfila Wildlife Reserve in northwest Spain, where they observed mothers feeding their male chicks an average of sixty-four times per day, whereas female chicks were fed only fifty-three times, a differential of 20 percent.[32] It may be that male chicks get more supplemental feedings simply because they beg more persistently, but it also makes sense for mothers to feed their sons as much as possible because there is no profit in rearing a male chick that will grow up to be too small to be competitive in the breeding hierarchy. Well-fed male chicks have high growth rates and are the first to become independent and disperse away from their natal areas. They also settle on a breeding lek at a younger age, become established in the male dominance hierarchy earlier, and have a higher chance of becoming a dominant, breeding male.

Given these benefits of rapid early growth, one can see why mothers endeavor to feed their male chicks as much as possible. However, the strong dependence of male fitness on rapid early growth comes at the cost of increased mortality, especially in years of low food abundance.[33] It appears that males gamble for the benefits of rapid growth and large size at maturity at risk of early death. By comparison, females adopt a more conservative strategy of slower growth and smaller adult size in return for a higher probability of survival to reproductive age.

Sex-specific life histories continue to be manifest during the years between fledging and sexual maturity. Males become independent of their mothers several months before females (table 4.1), and shortly thereafter they leave the security and familiarity of their natal foraging grounds, flying to distant areas where they join with other juvenile males in foraging flocks. Females tend to stay in their natal flocks longer than males, and if they do disperse, they tend to travel less than a third as far as males. Even if females move away as juveniles, they tend to return to their natal areas when they reach sexual maturity and generally visit leks and mate within a few kilometers of their natal site. Males, in contrast, typically remain far from their natal sites and eventually settle on leks up to 65 km (40 miles) distant.[34] Once they are mature, males and females continue to have different movement patterns and are seldom in the same place at the same time.[35] In Spanish populations most males migrate away from the leks in late May and early June immediately after the breeding season. They move an average of 82 km (51 miles), many flying to cooler, summering areas at higher altitudes and latitudes. In early fall these migrants either fly back to the lekking areas directly or detour to visit foraging areas that overlap those used by migratory females. By comparison, the adult females are much more sedentary. Almost half, including all females with chicks, remain close to the nesting areas throughout the year. Those that do migrate do not leave until fall or early winter, when they fly to alternate foraging areas, moving an average of only 50 km (31 miles).

The migratory patterns of great bustards in more northerly populations provide an interesting contrast to those in the Spanish populations. In northeastern Europe and central Asia bustards have a typical temperate migration strategy with both sexes migrating long distances in autumn to avoid the harsh winter weather. However, in central Europe where the climate is somewhat milder, winter migration is facultative and only occurs when persistent snow cover prevents the birds from feeding for three or four weeks. When this happens many fly up to 650 km (404 miles) west and south until they reach snow-free areas that offer adequate foraging. As in the Spanish populations, the likelihood of migration depends on sex, but here the females are more likely to migrate than the males. These differences in migratory patterns between sexes and regions probably derive from size-related differences in the energy budgets of males and females. Spanish researchers have proposed that males in their populations migrate to cooler areas in summer because their large body mass makes it difficult for them to thermoregulate in the high summer temperatures of the lowland areas. By comparison, German researchers have proposed that males in central European are less likely to migrate to warmer climes than females because their larger size allows them to tolerate longer periods of food shortage in cold weather. The larger size of the males also makes long migratory flights more energetically costly, and the net result is that the energy balance for males is likely to tip in favor of waiting out the snowy season in the breeding area, whereas for females it favors temporary migration to a warmer, snow-free foraging area. According to these arguments the sex-specific migratory patterns in both regions are indirect consequences of the large sexual size dimorphism of the adult birds.[36]

We have now come full circle in the lives of male and female great bustards. They are unusual animals in many ways: the largest of flying birds, the most sexually dimorphic of all birds in terms of body size, and a composite of the behavioral, morphological, and ecological characteristics of cranes and grouse. As we have seen, with the exception of brief encounters on the lekking grounds, males and females lead separate lives year-round. They disperse and populate the landscape independently, at different times, and to different degrees, and they respond to seasonal variations in temperature and food availability in ways at least partially dictated by their differences in body size. Females focus on rearing chicks, whereas males focus on mating with as many females as possible. Success on the lek is the focus of every male's life history, and lekking occupies a major part of each year for sexually mature males. In contrast, females visit the lek only briefly and seem to mate only once each year. If they are successful in rearing a chick, their parental duties occupy them almost year round, and this is the focus of their life histories. Although females spend little time on the lek, the lekking system benefits them by providing one-stop shopping for mates and hence reducing the risk, time, and energy devoted to finding a mate. More significantly, it allows females to readily compare males and to mate preferentially with the males in the best condition and with the flashiest displays. This imposes a high cost of sexual selection on males, but it benefits females by ensuring that they mate with only healthy males and that the resulting offspring receive only high-quality paternal genes.

Lekking is not a common mating system in animals but is relatively widespread in vertebrates and insects.[37] It is especially well-known in gallinaceous birds (the gamebirds) such as prairie chickens, sage grouse, black grouse, and peafowl, and is also characteristic of the gorgeous birds of paradise from New Guinea and Australia and the equally gorgeous manakins of the neotropics. Choruses of male bull frogs around a pond, clusters of male fruit flies on a bracket fungus, and swarms of male midges hovering in a patch of sun are other familiar examples of leks. Wherever they occur lek mating systems are associated with very high variance in male mating success and very strong sexual selection on males. Typically, in any given mating season, 60 percent to 90 percent of males on the lek fail to mate at all.[38] The result of this intense sexual selection is that males in lekking species have evolved exaggerated characteristics that enhance their mating success. Bright colors and vigorous behavioral displays are typical, as are loud calls in many species (but not great bustards).[39] In species that display and mate on the ground, large size and visible weaponry are also often favored.

There are many parallels between the typical pattern of reproductive role division in lekking species, as illustrated by great bustards, and the situation in elephant seals. In both cases males initially compete with each other through physical contests to establish a dominance hierarchy on the mating grounds, and they begin this process before the females arrive. Only males high up in the hierarchy subsequently get to mate, and these are generally larger, older, more aggressive males in top physical condition. Once females begin arriving the focus shifts from battling with other males to mating with as many females as possible. It is only during this phase of the mating cycle that the two mating systems diverge. Male elephant seals compete for mates by defending groups of relatively sedentary, spatially aggregated females from the advances of other males. In contrast, male great bustards compete by trying to attract mobile, choosy females that make only brief visits to the breeding grounds. Male elephant seals are able to force copulations on reluctant females and do so repeatedly. In contrast, male bustards must persuade choosy females to allow them to copulate. Sexual selection now favors different traits in the two species. Great bustard males rely on gorgeous visual displays and persistent courtship, whereas elephant seal males continue to rely on brute force. As different as these strategies may seem, however, both require health, endurance, and physical stamina, and only males that begin the mating season in top physical condition can succeed. In both species the requirements for mating success are stringent, and the majority of males in any generation never succeed in fathering offspring. To make the grade a male has to grow rapidly when he is young, and he must continue to accumulate body weight and fat reserves once he has become sexually mature. To do this he must be successful at foraging, escape debilitating disease or parasitism, and avoid death by accident or predation.

The parallels between great bustards and elephant seals are also obvious in female life histories. In both species females reproduce annually and typically rear only one offspring per year. They maximize their lifetime reproductive success by maturing at a younger age and smaller size than males and by returning to mate and breed for as many years as possible. To ensure that their offspring grow rapidly and survive to independence, female elephant seals sacrifice their own energy reserves to provide nutrition for their offspring. Perhaps because of this focus on offspring production and survival, female elephant seals lead much more conservative lives than males. Their interactions with other females are less aggressive than those of the males and are seldom harmful, and they face much less variance in reproductive success than males do. It is only in their interactions with males that females of the two species differ dramatically. Female elephant seals are clearly in sexual conflict with males and frequently endure unwelcome copulation attempts that are occasionally injurious. Males control a resource that the females need—safe beaches for breeding—and females that haul out to give birth cannot avoid harassment by males. Great bustard females have none of this. Males do not control any resources needed by females and instead must entice females to mate. This puts females in control of copulations and greatly increases the opportunity for sexual selection through female choice. The result is that the same males that have been tearing each other up in aggressive contests for dominance on the lek elegantly strut and puff themselves up when females approach. This is quite a contrast to the brutish mating squabbles in the elephant seal rookery. In the next chapter we meet a species in which males defend resources that females need but also rely on the vagaries of female choice to achieve matings. Here the importance of female reproductive tactics as determinants of sexual differences in body size will become even more apparent.

CHAPTER 5

Shell-Carrying Cichlids

Protective Males and Furtive Females

Elephant seals and great bustards represent the extremes of sexual size dimorphism within mammals and birds, and the differences between the sexes in these species are certainly impressive. However, to find extraordinary differences between males and females among vertebrate species, we have to leave the familiar world of large, terrestrial animals and explore the watery lifestyles of fishes. The ray-finned fishes (class Actinopterygii) are the clear vertebrate champions in terms of both the diversity and the magnitude of their sexual differences. At one extreme are species in which females can be hundreds of thousands of times more massive than their mates. This pattern of dwarf males and giant females has evolved independently in a number of groups of deep- and midwater marine fishes, and we have a close look at one of these, the giant seadevil, in chapter 8. In this chapter we meet a small, freshwater fish at the other extreme of sexual dimorphism, a species in which males typically weigh almost thirteen times more than females. This tiny fish, called *Lamprologus callipterus* (plate 11), is the runaway (or rather, swim-away) champion of male-larger sexual size dimorphism not only among fishes, but across the entire animal kingdom.

Lamprologus callipterus belongs to a large and diverse family of perchlike fishes called the Cichlidae, and it is one of the more than 300 species of cichlids found only in Lake Tanganyika in Africa.[1] Males are the larger sex in most African cichlids, including most *Lamprologus* species, so the presence of male-larger sexual dimorphism in *L. callipterus* is not unusual. What *is* unusual is the magnitude of this dimorphism. In other African cichlids males range from 70 percent longer to 16 percent shorter than females, whereas *L. callipterus* males average 2.4 times (140 percent) longer than their mates.[2] This difference between the sexes is even more pronounced for body mass: the average weights of breeding male and female *L. callipterus* are 26.7 g and 2.1 g respectively, giving an average mass ratio of 12.5.[3] Surprisingly, unlike male elephant seals and great bustards, *L. callipterus* males are not distinguished by being the largest males in their family or even in their genus. Compared to other *Lamprologus* males they are only intermediate in size at best (figure 5.1). The unusual feature of *L. callipterus* is the small size of the females, which are the smallest in their genus, rather than the large size of the males. Male *L. callipterus* are not large when compared to males in related species, but they are truly giants relative to their females, and this is what generates the extraordinary sexual size dimorphism in this species.

The key to understanding this sexual dimorphism lies, as you may have guessed, in the reproductive behavior of males and females. These little fishes spawn and brood their eggs within empty snail shells on the lake bottom, a breeding strategy that has evolved repeatedly in African cichlids.[4] In all other shell-brooding cichlids, however, both sexes enter the shell to spawn, and so both have to be small enough to fit into the shells available on the lake bottom. The key difference in *L. callipterus* is that only the females enter the shell to spawn and brood their eggs. The males remain outside, releasing their sperm only at the entrance to the shell. As a result female *L. callipterus* are small enough to fit into the shells and similar in size to other shell-brooding cichlids (all of which are in other genera), but the males, unconstrained by the need to enter shells, are much larger. As we shall see, the large size of the males is the result of intense sexual selection combined with their unique behavior of actually picking up and carrying shells to serve as brooding chambers for their mates.[5]

A good place to begin our *L. callipterus* story is at the start of the female spawning cycle, when females begin their search for suitable shells

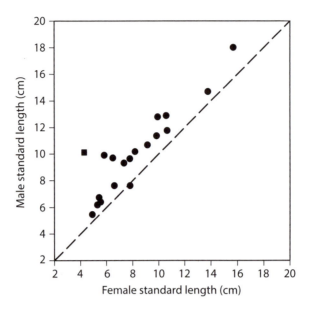

FIGURE 5.1. Mean standard lengths for fishes in the genus *Lamprologus*. Male lengths are plotted on the vertical axis, and females on the horizontal axis. The dashed line is the line of equality. Points above that line signify male-larger sexual size dimorphism. Data are from Ota et al. (2010) for *Lamprologus callipterus* (■) and from Erlandsson and Ribbink (1997) for the other species (•).

to shelter their broods. When they are not spawning or brooding, female *L. callipterus* swim in mixed shoals of males and females in the shallow, inshore areas of the lake feeding and building up their energy reserves while they mature their next batch of eggs. Once a female's eggs are mature and ready to be fertilized, she leaves the shoal and sets about finding an empty snail shell large enough to serve as a spawning and brooding chamber. With this objective she cruises over the bottom searching for and inspecting snail shells in the appropriate size range. To spawn she must be able to fully enter the shell, turn around, and glue her eggs to the inner side of the shell whorl. The eggs hatch in three to four days, but the tiny larvae (called wrigglers) remain inside the shell for about two more weeks until they have absorbed their yolk sacs and are ready to swim away as independent fry. Their mother stays inside the shell with them through this entire period, flicking her tail to aerate the brood and guarding them from predators. Females typically lay between 50 and 250 eggs in the shell, and about 70 percent of these survive to independence.[6] This means that to serve as a brooding chamber, the shell has to accommodate not only the 4 cm-long mother, but also 35 to 175 squirming wrigglers, each of which will be about 0.7 cm long by the time it abandons the shell. Not surprisingly, finding a snail shell large enough to suit and that is not already occupied by another female can be a daunting task for a female bursting with ripe eggs.

In laboratory trials females generally choose the largest shells available, and it seems likely that they have the same preference in the wild.[7] Larger shells are superior because they can accommodate larger broods and also allow more efficient aeration of the eggs and fry. A larger shell also permits the female to lay her eggs deeper within the whorl and to station herself more deeply within the shell to protect her brood from predators. This is a key advantage because egg predation by other fishes, including members of her own species, is common. The problem for females is that large shells can be hard to find and large, empty shells even more so. The distribution of snail shells is very patchy, and both shell size and shell abundance vary a lot from place to place. At least three snail species have shells that can be used, but female *L. callipterus* much prefer the shells of the largest species, *Neothauma tanganicense*. In some areas of the lake these coveted shells are relatively rare, and competition for them is intense. Females typically search for about two days before securing a suitable shell, and, once they find one, they defend it aggressively against females of their own and other species. In this competition larger females generally win, with the result that the largest females typically secure the largest shells. The other females sort themselves out down the line such that, on average, each ends up spawning in a shell slightly larger than her own body size—but probably not as large as she would prefer. Unfortunate females that end up spawning in shells that are too small have a high risk of failure. Many lose their broods to predators, and others simply abandon the shell when their brood begins to take up too much space. These failures have severe consequences for the mothers because it will take them many weeks to mature another batch of eggs and spawn again. Thus, the cost of failing to find a sufficiently large shell is high in terms of both loss of the current brood and delayed reproduction.[8] It is not surprising that females search diligently to find large shells and that they aggressively defend a good shell once they have found it.

Shell brooding is a very effective strategy for rearing offspring in the predator-dense environment of the shallow lake bottom. However, spawning and brooding inside shells puts females in an evolutionary double bind with respect to their body size. In cichlids, as in most fishes, the number of eggs spawned per batch (the *fecundity*) increases with female body size, and this favors females that grow large. In *L. callipterus* the correlation between female length and number of eggs spawned is as high as 0.95, which means that more than 90 percent of the variation in fecundity among females can be explained by differences in body length.[9] Further, because fecundity is more a function of body volume than length, it increases disproportionally as body length increases.[10] In one study, for example, an increase in female length from 3 cm to 4 cm was associated with an increase in fecundity from 59 to 134 eggs. This is a 227 percent increase in fecundity for a 33 percent increase in body length.[11] By the time the young had reached independence, the 3- and 4-cm long females had 30 and 104 fry respectively, a differential of 347 percent! In addition to this huge fecundity boost larger females are able to exclude smaller females from access to the larger, preferred shells and so gain in the competition for spawning sites. All this suggests that female *L. callipterus* should be large, but this is where the double bind comes in. No matter how beneficial large size might be, the paucity of large shells ultimately limits the size that females attain in a given population.[12] In laboratory experiments females actually adjust their growth rates so that their body size at maturity matches that of the available shells, and in the wild the correlation between shell size and female size explains more than 98 percent of the variation in female size among populations. The ultimate size of female *L. callipterus* is thus a compromise between growing as large as possible to maximize fecundity and remaining small enough to be able to find suitable shells for spawning and brooding their young.

Females follow a life history that reflects this compromise.[13] They reach sexual maturity quickly, at about six months of age, by which time they have typically reached standard lengths of 3.5 to 4.5 cm and weights of at least 2 g. Their growth then slows dramatically, and although they may live as long as three years in the wild, they rarely reach lengths exceeding 6 cm or weights exceeding 5.5 g. Throughout their adult life they alternate between shell-brooding, which takes about two weeks, and extended periods of foraging, which last at least seven weeks. Because fewer than one third of broods survive to independence, the females have to rely on repeated spawning attempts and a long reproductive lifespan to achieve high fitness. To that end they begin reproduction at an early age and reduce their risk of mortality by remaining cryptically colored throughout their lives. Surprisingly, one of the main threats to female fitness is the behavior of their own males (more on this later), and even a female's own mate is likely to eat eggs or larvae if they are left unguarded in the shell. A female in her shell has no friends or allies. Should she choose a shell that is too small or otherwise fail to protect her brood, she has no recourse but to return to foraging until she has recouped her energy and produced another clutch of eggs for spawning.

Finding shells large enough for brooding is so critical for female *L. callipterus* that, when they are ready to spawn, they generally pay more attention to choosing shells than to choosing mates. Males have adopted a unique strategy to exploit this female choosiness.[14] Rather than trying to attract females from afar with flamboyant displays, male *L. callipterus* entice females to mate with them by assembling piles of empty snail shells. They set up small territories, less than a meter in diameter, and each territory is centered on a patch of shells. In most areas of the lake these patches of shells are separated by stretches of sandy or rocky bottom devoid of shells (plate 12). The males create this patchy distribution by picking up and carrying shells from as far away as 20 m. They manage this by clamping their mouths on the open edge of the shell, lifting it off the bottom and swimming with it back to their pile. In addition to clearing the surrounding area of abandoned shells, territorial males frequently steal shells from the shell piles of other males, sometimes with females and brood inside. When this happens the female and her brood usually disappear, and the empty shell becomes available for a new female in the poacher's patch. The effect of this remarkable shell transporting is to create artificial clumps of shells surrounded by unsuitable stretches of lake bottom. When females are ready to spawn, they have to visit the shell piles of territorial males to find suitable shells, and this creates spawning opportunities for the males.

Males spend most of their time defending and augmenting their shell piles, but if a female ripe with eggs approaches, the resident male quickly switches to courting mode.[15] He typically approaches her and performs a series of vigorous, zigzag turns to catch her attention. As she examines a shell, he may jerk his head back and forth in front of her and repeatedly mouth the shell as though pointing out its desirability as a brooding site. Enthusiastic males may even push and gently mouth the female to encourage her to enter the shell. If the female is satisfied with the shell and its owner, she will curl inside the shell and prepare to spawn. The male assesses her readiness by gently mouthing the entrance to the shell while undulating his body. When she is ready she flicks her caudal fin, drawing attention to her swollen belly, and then deposits one egg on the inner whorl of the shell. Since the male cannot enter the shell, he fertilizes this egg by placing his genital papilla over the shell entrance for several seconds and ejaculating, trusting that his sperm will find their own way to the inner whorl where the egg lies. The female waits patiently inside the shell as he does this. The couple repeat this sequence of oviposition and ejaculation for up to twelve hours, until the female has spawned her entire clutch. The male then leaves her ensconced in the shell with her eggs while he returns to guard his shell pile and entice other females to spawn with him.

In established populations most males initially acquire shell piles by taking them over from other territorial males.[16] Understandably, males vigorously defend their shell piles against such threats. If another male intrudes the territorial male faces him, tilts his head down to display his colorful dorsal fin, and undulates his body. This is often sufficient to discourage the intruder, but if not, chases and physical fights ensue, and these can last several days. Researcher Tetsu Sato[9] observed one such contest in which a larger male persistently attacked a smaller territory holder, attempting to bite him up to thirty-seven times in a single ten-minute period. These attacks continued sporadically day after day. The intruder also harassed the females brooding in the territory by picking up and shaking their shells, often causing the brood to fall out of the shell where it was generally gobbled up by the bullying male. Finally, after nine days of embattlement the original territory holder disappeared, ceding his shell pile to the larger male. In these wars of attrition victory most often goes to the larger male, and this is one of several ways by which larger males are favored in the intense competition for mates.

To successfully reproduce, a territorial male must not only acquire a shell pile, he must also hold on to it long enough for his mates to rear their broods. If he is deposed or deserts his shells, the new territory holder will quickly expel most of the brooding females and their broods to make the shells available for new females that will spawn with him.[17] The longer a male can hold his territory, the more of his offspring will safely mature. By holding his territory longer the male is also likely to acquire more mates because new females are continually arriving to inspect the shells. In the stable, tropical climate there is no natural end to the breeding season that would free males from their territorial battles and allow them to replenish their energy stores. Instead, territorial males simply hold on as long as they can. Because they eat little while courting, spawning, and defending their shells, their condition declines day by day. Ultimately they are either displaced by challengers in better condition or simply abandon their territory when their reserves become too depleted. They then join the shoals of males and females foraging along the bottom for shrimp and the eggs and larvae of their own and other species, and they do not attempt to secure another territory until their condition is restored.

The average duration of territory tenure is about a month but can be up to four months for the largest males. Just as we saw in elephant seals, and for similar reasons, larger males are able to hold their territories longer than smaller males and hence have a reproductive advantage. This is partly because larger males have greater initial energy reserves and lower mass-specific metabolic rates than smaller males, and so their reserves last longer.[18] As in elephant seals and great bustards, larger *L. callipterus* males have a behavioral advantage as well. They are more likely to win territorial disputes and hence less likely to be deposed before they are ready to leave. The net effect of this is that, even in populations where females do not seem to choose their mates based on body size, larger males accrue a significant mating advantage through their success in acquiring and holding mating territories.

These advantages alone would probably be sufficient to cause males to be larger than females, especially given that females are constrained to be small enough to brood their offspring inside a snail shell. However, the extreme sexual size dimorphism shown by *L. callipterus* requires additional explanation. The clue for this is the unique shell-carrying behavior of this species. The males have to be large enough to hold, lift, and carry snail shells suitable for use as brooding substrate. The *Neothauma* shells preferred by females are quite large, with volumes typically exceeding 15 cm³ and heights of 4–6 cm. Carrying them is no small feat for a fish whose own body length is only in the range of 9 to 14 cm.[19] Males smaller than about 9 cm are unable to lift these large shells off the bottom and so cannot transport them at all. This threshold is well above the largest sizes attained by females (about 6 cm), and so it establishes a baseline sexual size dimorphism imposed by the need for one sex to fit into the shell and other sex to carry it. Above the minimum threshold the efficiency of carrying shells increases disproportionally with body size such that larger males can carry shells that are heavier relative to their own weight. Selection for efficient shell carrying acts as an indirect form of sexual selection on males because a male's success in accumulating large shells determines his potential mating success.[20] Larger males can carry larger shells and transport them more efficiently, so they are able to accumulate more shells and larger shells in their shell piles. This allows them to attract more females and achieve higher mating success than their smaller competitors. Larger shells can also accommodate larger females that will have higher fecundity, so larger males are likely to have the added benefit of more surviving fry per mating.

Male *L. callipterus* are using a mating strategy that biologists call *resource defense polygyny*. The number of females that each male attracts (i.e., the extent of polygyny) is proportional to the number of desirable shells in his pile and depends on how long he is able to defend this pile against thieves and usurpers. Rather than devoting himself to attracting females from afar by elaborate displays or to solicitous care of his mates and offspring, he focuses on accumulating more and larger shells and defending these against other males that attempt to steal shells, usurp his territory, or simply prey on the eggs and wrigglers. The number of females a male has brooding in his shell pile at any given time typically ranges from two to six but can be as high as thirty, and the most successful males may spawn with as many as eighty-six different females over a single territorial tenure.[21] The net effect of this mating strategy is strong selection for large size in males. Larger males have territories containing more and larger shells, and they keep these territories longer. As a result, they mate with more females, fertilize more eggs, and father more fry than smaller males (figure 5.2).

Males on the trajectory to become territory holders have a very different life history than females. They grow twice as fast during their juvenile phase but do not mature until about a year later than females. By that time they are more than 9 cm long and weigh more than 20 g, about two and a half times as long and ten times the weight of females at maturity. This gap between the sexes continues to widen with age because males continue to grow at a faster rate than females throughout their adult lives. No explicit comparisons have been made between male and female lifespans, but males continue to grow for at least thirty-three months in the laboratory and are known to live for up to three years in the wild. Like females, territorial males alternate between breeding and foraging, but successful males may spend as long as four months at a time on their territories, much longer than the two-week stints of females. Rather than being small, cryptic, and reclusive like their mates, the territorial males are large, active, and flamboyant. They challenge and chase rival males and investigate any ripe females visiting their territories. They jerk and tug and prod to encourage females to choose shells and to spawn, and

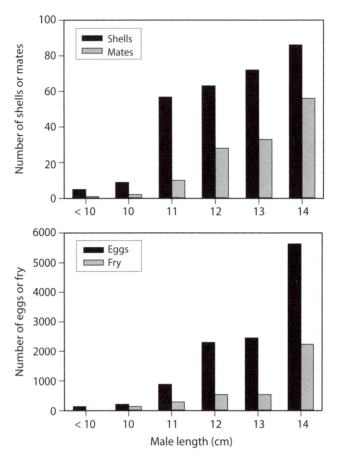

FIGURE 5.2. The relationship between body size and reproductive success for territorial male *L. callipterus*. The top panel shows the average numbers of shells (black bars) and females breeding (gray bars) in the shell piles of males in six size classes. The bottom panel shows the average numbers of eggs spawned (black bars) and young reared to independence (fry; gray bars) for the same male size classes. Size is measured as standard length. Data are from Sato (1994).

then they ejaculate again and again to fertilize each egg laid. They spend about 15 percent of their time courting and 10 percent defending their territories, and when not busy with aggression, courtship, or spawning, they are likely to be transporting and rearranging their shells, or stealing shells from neighbors. All in all they are actively moving about 35 percent of the time. While the brooding females are hiding quietly in their shells, the territorial males are constantly alert and exposed in the open water, even when resting. The behavior, size, and appearance of the two sexes are so different that one would never place them in the same species if they were not obviously mating partners.

The success of resource defense as a mating strategy for males and hence the strength of selection favoring large males depend on the sizes and relative abundance of shells on the lake bottom.[22] The males are largest and mating success is most skewed in populations where the large *Neothauma* shells are available but in short supply. This is the most common situation and explains the pattern of sexual dimorphism found in most areas of the lake, with territorial males averaging twelve to thirteen times heavier than females. However, there is one area where *N. tanganicense* shells are absent, and *L. callipterus* use the much smaller shells of another snail species. Female size matches the small size of the available shells, and they are the smallest breeding females in the lake. However, the extreme polygyny in this population still favors large males, so male size responds less strongly to the small size of the shells. The result is by far the most extreme sexual dimorphism in the lake, with territorial males averaging sixty times heavier than their females.

The reproductive behavior of territorial male *L. callipterus* is clearly analogous to the harem master tactic of elephant seals. The objective of this tactic is to obtain matings by monopolizing a resource needed by females (shells or beach area) and keeping other males away. Only large males in top condition can succeed via this tactic, and this imposes strong selection for large male size in both species. For males unable to acquire a harem or territory, the only option is to try to sneak or force copulations when the territorial male or harem master is otherwise occupied. To this end males often hang around on the periphery of shell piles and dash in to quickly court and fertilize unguarded females when the guarding male is busy elsewhere.[16] Males that do this are called "sneakers," and they are usually smaller and younger than the territorial males. They employ an opportunistic and transitional tactic analogous to the peripheral males in an elephant seal colony. Sneakers are vigorously attacked by territorial males if caught, and they seldom achieve fertilization success. However, given that a male might not survive long enough to gain a territory, sneaking is a good tactic as long as it remains low risk, does not detract too much from foraging, and does not require much energy. If these criteria are satisfied, sneaker males can continue to eat and grow, preparing themselves for their future as territorial males, while possibly fathering a few offspring on the way.

Whereas some males adopt the sneaker tactic on their way to becoming territorial males, others commit themselves much more fully to the tactic of gaining fertilizations by stealth. These males abandon the growth trajectory that could lead to territorial status and instead mature at least a year earlier, when they are as small as 2.4 cm long and weigh as little as 0.3 g. Unlike the sneaker tactic this life history is genetically based, and these males are dwarfs for life.[16,23] They grow more slowly as juveniles and stop growing entirely by the time they are two years of age, whereas males on the territorial track grow faster initially and continue to grow throughout their lives.[24] The dwarf males are noticeably smaller than mature females but otherwise look so much like females that they have been called female mimics.[9] No doubt this camouflage aids in their subterfuge as they try to slip into shells occupied by ripe females without being detected by the guarding male. Once inside, unbeknownst to the territorial male, they ejaculate and generally succeed in fertilizing the majority of eggs as they are spawned. To be successful the dwarfs need large testes to produce copious amounts of sperm. When mature, their testes comprise up to 2.0 percent of their body weight, whereas the comparable value for territorial males is less than 0.4 percent. Unlike territorial males, which eat little or nothing while defending their shell piles, dwarf males spend about 20 percent of their time foraging around the periphery of the territories. Because they are not growing, carrying shells or defending a territory, they are able to allocate almost all of the resources they gain from feeding to sperm production and mating attempts. Thus, while the territorial males experience a continual decline in condition and eventually have to abandon their shell piles and return to foraging, the dwarf males can maintain a steady rate of subversive mating attempts throughout their adult lives.

The dedication of the dwarf males to achieving subversive matings is quite astonishing. They typically try to mate ten to eleven times each day, darting over the piles of shells, on the lookout for ripe females while at the same time trying to avoid the guarding males. Sometimes they behave like sneaker males and attempt to quickly court and spawn with unguarded females, but these attempts are invariably unsuccessful. Their only profitable tactic is to wriggle inside a shell either before or while the territorial male is courting and spawning. Not surprisingly, territorial males vigorously defend their females against these intrusions by dwarfs and chase most of them away before they can enter the female's shell. The dwarfs may even be injured or killed by the territorial male in these interactions, so a mating dash is not without risk. In fact the dwarf tactic seems to be a high-risk and generally marginal reproductive tactic for *L. callipterus* males. Being a dwarf works only in populations where the brooding shells are large enough to allow the tiny dwarf to slip into the shell past the ripe female and to remain in the shell long enough to fertilize the female's eggs (often more than six hours). Since most of a dwarf male's mating attempts will be failures, the tactic also requires a high density of ripe females. Even where these conditions prevail, the territorial male generally succeeds in fertilizing more than 90 percent of the eggs spawned in his territory. Not surprisingly, dwarfs are often rare or absent entirely and represent a distinctly minority tactic. However, where they do occur, they provide a striking example of male alternative reproductive tactics and one of relatively few cases where such tactics have been shown to be genetically based and fixed for life.[25]

L. callipterus have become quite notorious for their extreme sexual dimorphism, which of course refers to the size of the territorial males relative to their mates. These males are huge compared to breeding females, and it is not surprising that researchers have wondered why the sexes differ to such a degree. In one respect the *L. callipterus* story has turned out to be exceptional. This is the only known species of shell-brooding fish in which males actively transport shells by carrying them through the water, and the need for males to carry the shells in which their mates will shelter certainly favors males that are much larger than females. However, although adaptation for carrying shells is unique to *L. callipterus*, the other aspects of their story resonate with what we have seen in elephant seals and great bustards. In all three species sexual selection favors large size, aggression, and attention-grabbing displays in males, whereas reproductive trade-offs favor smaller size, less aggression, and more cryptic behavior in females. Sexual selection is particularly strong in these species because reproductive adults congregate in established areas for mating, which fosters direct and intense competition among males for mates. In addition none of the males provision their mates or offspring, and so they are free to devote themselves to the job of mate acquisition rather than to parental care.[26] Equally significant is that females in all three species provide postnatal care for their offspring, and this limits both the number of offspring they can rear in a given brood and their own potential for growth. To achieve high lifetime fecundity females have to raise multiple broods, and all three species are characterized by long reproductive lives with multiple breeding episodes separated by long nonreproductive periods. All three species are also to some extent *capital breeders,* which means that they do not eat during at least some phases of the breeding episode. The necessity of fasting while breeding contributes to their extreme dimorphism by increasing the impact of the trade-offs among reproduction, growth, and body maintenance in both sexes. As we have seen, the net result of these trade-offs is smaller size in females but larger size in males.

Although these are only three examples their similarities suggest a suite of characteristics that are conducive for the evolution of pronounced male-larger sexual dimorphism. When all of these characteristics are in place, males and females carve out very different life histories for themselves. Females follow a relatively low-risk, conservative life history in which they grow more slowly, begin breeding at an earlier age and smaller size than males, and behave in ways that maximize their survival and that of their dependent young. Once mature, they reproduce at a relatively low rate but over many consecutive breeding episodes, allocating much of their energy to postnatal brood care. They move inconspicuously through their world, eschewing flamboyant displays or conspicuous coloration, and they engage in much less intense aggressive behavior than their mates, unless in direct defense of their young. Males, in contrast, grow faster and mature at a later age and much larger size than females. They focus on out-competing other males for access to mating opportunities, and this generally entails a great deal of aggression and self-advertisement. They pay no heed to their offspring and are more likely to be a threat than a help to brooding females. At best they may protect their mates and off-spring from harassment by rival males or predators, but great bustards do not even do that. This is a high-stakes, high-risk life history where bru-tal fights between males are not uncommon, and everything depends on being better than the other males. Social interactions between the sexes are confined almost entirely to mating, and sexual conflict rather than cooperation is the norm.

These, of course, are examples of extraordinary differences between the sexes. Nevertheless, aside from the shell-carrying behavior of *L. cal-lipterus*, there is nothing qualitatively different between these three spe-cies and many other species of animals where males are polygynous and larger than females and females provide postnatal parental care. This is ac-tually a common way of partitioning male and female reproductive roles, especially in vertebrates. As a general rule in animal lineages where this pattern holds, sexual dimorphism increases with the strength of sexual selection on males.[27] The positive relationship between sexual selection and sexual dimorphism suggests that sexual selection is the primary cause of the differences between the sexes. However, I would argue that selec-tion on females is equally important. In the clades where sexual selection seems to have such a strong effect, females remain relatively small and inconspicuous because increasing size would reduce their lifetime repro-ductive success. Larger females must invest more energy in growth and body maintenance rather than offspring care, and increasing conspicu-ousness increases the risk of mortality for both mothers and their off-spring. The net effect is that larger or more conspicuous females would produce fewer offspring. Thus, although sexual selection has caused their mates to become large and conspicuous, females remain smaller and more cryptic because this is what maximizes their Darwinian fitness. Selection keeps females small and inconspicuous to the same extent that it makes males large and flamboyant. As we see in the following chapters, when the pattern of selection on females changes, the effect on sexual dimorphism can be dramatic. In particular when female reproductive success depends primarily on the size of individual clutches rather then success in multiple breeding episodes, and when mothers do not care for their offspring after birth or hatching, females are generally larger than their mates. This can be true even if large males have an advantage in competition for mates, as we discover in the next chapter where we explore the extreme sexual dimorphism in orb-web spiders.

CHAPTER 6

Yellow Garden Spiders

Sedentary Females and Roving Males

My first encounter with a yellow garden spider occurred quite unexpectedly on a cool September morning as I was clambering up a sandy bank on the edge of an abandoned quarry in southern Quebec. I had been looking down as I scrabbled for handholds on the slippery bank, and I glanced up just as I crested the lip of the quarry. The low morning sun dazzled my eyes, glancing off a fine layer of dew that still lingered on the tangle of early fall herbage. In the midst of this chaotic brilliance, suspended no more than a hand's breath from my face, was an enormous spider. She was hanging head downward in the middle of a beautiful orb web, every strand of which was sparkling with water droplets. The web was at least 50 cm (20 inches) in diameter, and the spider's body looked as big as the end of my thumb. She showed no indication of noticing me even though I came perilously close to blundering into her web and destroying it entirely. I came to an abrupt stop, jerking back in response to a primal aversion to close facial contact with spiders or their sticky webs. Having grown up in eastern Canada I was moderately well acquainted with the various orb-web spiders there. In my experience they were rather drab beasts, sporting subdued patterns of brown and tan, but this spider was something else entirely. Not only was she by far the largest spider I had ever seen, she was also dramatically striped in yellow and black and was remarkably conspicuous. As though to draw even more attention to herself, she had constructed a wide zigzag of white silk through the center of her web. The orb-web spiders I normally encountered hid off to the side of their webs, no doubt to make their webs and themselves less conspicuous. This spider was clearly adopting a different strategy, hanging audaciously in the middle of her web in the bright morning sun.

In my university office later that day I easily identified my mystery spider as *Argiope aurantia*, the yellow garden spider (plate 13). According to my field guide these conspicuous black and yellow orb-weavers can be found from southern Canada to Costa Rica and throughout the contiguous United States. The species was new to me simply because I had been living in areas that were very slightly north of its range. As their name suggests, garden spiders often build their webs in sunny locations in backyard gardens and old fields. However, their habitat choices are actually quite catholic and include almost any habitat where shrubs and tall herbaceous vegetation occur in sufficient densities to permit web construction.

Having thus satisfied my curiosity I thought little more about garden spiders until a graduate student named Matthias Foellmer joined my laboratory with the intent of studying sexual dimorphism in spiders. I was a professor at Concordia University in Montreal at the time, and it made sense for Matthias to study a species common in our local habitats. He prowled the roadsides and old fields within driving distance from campus seeking a species that showed marked sexual dimorphism and was abundant enough that he could obtain good sample sizes for his field studies. He found good populations of *A. aurantia* in several old fields and, to my delight, chose this large, conspicuous species for his dissertation research. Much of what I discuss in this chapter comes from that research and the studies that have built on it.

Why would anyone choose to spend years studying sexual dimorphism in a spider? To most people such a project would seem arcane to say the least. For students of sexual differences, however, spiders have a special significance. They may well hold the key to unlocking one of the biggest mysteries remaining in our quest to understand sexual differences: why males are often so much smaller than females. Species in which males are

truly dwarfs relative to their mates occur in many different animal lineages, several of which we will meet in later chapters, but spiders are the
only group of readily observable terrestrial animals in which this pattern
of extreme sexual size dimorphism is common. The African golden orb
spider *Nephila turneri* holds the record among spiders, with female body
lengths almost ten times those of males,[1] but females are at least twice as
long as males in many species of orb-weaving and crab spiders. To discern the origins and adaptive significance of extreme sexual dimorphism
in these species, spider biologists have studied various aspects of spider
behavior, measured the sizes of thousands of males and females, and for-
mulated vast comparisons among species.[2] Matthias and I proposed to
join this fraternity by studying the full life histories of male and female
A. aurantia with the express aim of figuring out why the sexes are so different in this large and conspicuous species.

Yellow garden spiders are classified in the family Araneidae within the
large evolutionary lineage (*clade*) of "orb-weaving" or "orb-web" spiders
called the Orbiculariae.[3] Females are almost always the larger sex in this
clade, averaging about 2.3 times longer than males,[4] and females in the
genus *Argiope,* the garden spiders, average 3.5 times longer than males.
Our yellow garden spiders, *A. aurantia,* are among the largest and most
dimorphic with females reaching as much as 28 mm (1.1 inches) in some
populations and averaging four to six times longer than males (figure
6.1).[5] Because of the massive size of the female abdomens, especially
when full of eggs, the mass differential between sexes is much greater
than these length differences would suggest: mature females can weigh
as much as 1.5 g (0.05 oz), whereas males seldom exceed 20 mg (0.02 g),
and at the height of the breeding season females weigh an average of fifty-
three times more than males (table 6.1). Although certainly impressive,
A. aurantia is neither the largest nor the most sexually dimorphic spider
species. However, it is one of the few orb-web species in which researchers have attempted to measure selection through all phases of the life histories of both sexes, and it is the only species in which selection on leg
length and body size have been measured independently through each of
those phases. These details will become important as we try to understand

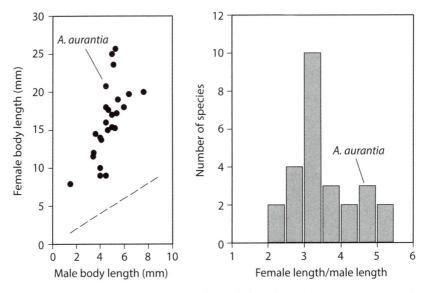

FIGURE 6.1. Sexual size dimorphism for body length in the genus *Argiope*. Left panel: Mean female body length versus mean male body length. The dashed line is the line of equality and points above this line signify that females average larger than males. Right panel: The distribution of ratios of mean female length to mean male length. The arrow indicates the position of *Argiope aurantia*, with an average length ratio of 4.6. Data are from Elgar (1991), Hormiga et al. (2000), and Wilder and Rypstra (2008).

why males are so very much smaller than females in this and other orb-web species.

The size differential is the most obvious difference between the sexes in *A. aurantia*, as in most orb-web spiders, but closer examination of the photo in plate 13 and the measurements in table 6.1 reveals noticeable differences in other characteristics as well. Males are slimmer and have proportionally smaller abdomens than females. Their legs are longer relative to their body size, and they have particularly long front legs. Males also have less dramatic coloration than females. Female abdomens are striped and spotted with bright yellow on a black background, sometimes with a pair of distinct white patches at the front. The other spider body component, called the cephalothorax or prosoma, is a combined

Table 6.1.

Characteristics of yellow garden spiders, *Argiope aurantia*

	Males	Females	Size ratio (female/male)	Ratio to width of prosoma Males	Ratio to width of prosoma Females	Sources
Body mass (mg)	16.1	847.1	52.7			1
Skeletal size (mm)						
Total length	3.5–5.5	19.5–22	4.0–5.6			2, 3
Width of prosoma[a]	1.8–1.9	3.5–3.7	1.91–1.95			4, 5, 8
Patella + tibia length[b]						
leg 1	3.84	5.99	1.56	2.16	1.71	4
leg 2	3.66	5.80	1.59	2.06	1.65	4
leg 3	1.75	3.28	1.88	0.98	0.93	4
leg 4	3.00	5.40	1.80	1.69	1.54	4
Age and growth						
Development time (days)[c]	45	59				5
Number of molts[d]	6–7	8–12				6, 7
Juvenile growth rate (mm/day)[e]	0.027	0.035				8

SOURCES: 1, Matthias Foellmer, unpublished data for wild spiders captured on August 25, 2009 on Long Island, NY; 2, Elgar (1991); 3, Hormiga et al. (2000); 4, Foellmer (2004); 5, Blanckenhorn et al. (2007); 6, Matthias Foellmer, unpublished data for laboratory-reared spiders; 7, Enders (1977); 8, Inkpen and Foellmer (2010).

NOTE: Values are means unless otherwise indicated.

[a] Spider bodies have two major parts, the prosoma (= head + thorax) and the abdomen.

[b] The patella and tibia are the fourth and fifth of the seven segments of spider legs. The legs are numbered from anterior to posterior.

[c] This is development time after emergence from the egg sac under laboratory conditions.

[d] Molts separate successive developmental stages (instars) that spiderlings go through as they grow. The final molt gives rise to the adult spider.

[e] Average increase in width of the prosoma per day between the second and the sixth molt under laboratory conditions.

head and thorax region sporting eight eyes and mouthparts on the front end, and four pairs of legs down the sides. In mature females the back of the prosoma is covered in a thick mat of silvery hairs, and the legs have distinct bands of yellow or orange with an especially broad band next to the body. Males lack the mat of silver hair and vary from medium brown to black with lighter areas on their legs and abdomens in similar but less distinct patterns than those of females.

These sexual differences are obvious even to untrained observers, but a true spider biologist would also note a somewhat more cryptic sexual dimorphism that is common to all spiders. On the front of the prosoma between the fangs and first pair of walking legs, both male and female spiders have a pair of appendages called pedipalps. During feeding, spiders use their pedipalps to touch, manipulate, dismember, and possibly taste the item being consumed.[6] The segment of the pedipalp closest to the body serves as a chewing mouthpart and is flattened with a strongly serrated cutting surface and hairs on the inside that probably act as strainers as the spider feeds. In females the remaining segments of the pedipalp, collectively called the *palp*, are thin and tapered and resemble delicate, diminutive legs. By contrast the palps of mature males resemble the arms of a pugilist, complete with lumpy, bulbous boxing gloves at the ends (plate 14). The end of the male's palp is a robust and complex copulatory organ composed of an expanded bulb that stores sperm, a coiled tube through which the sperm are transferred to the female, and a tapered, heavily sclerotized (hardened) tip called the *embolus* that serves as the intromittent organ (imagine a boxing glove with a curved pick sticking out the end). With the help of a hand lens or microscope, it is easy to distinguish mature males of any spider species, including *A. aurantia,* by the presence of these fully developed palpal organs poised between their mouths and first legs.

The distinct palps of male spiders are essential for sperm transfer, and they are precisely shaped to fit into paired genital openings in the female's *epigynum,* a slightly raised, sclerotized plate on the underside of the her abdomen. The genital openings lead to coiled insemination ducts that are also sclerotized so that only males whose emboli are the appropriate size and shape can achieve intromission.[7] In *A. aurantia* the tips or caps of the emboli have a complex shape resembling a scoop that has been curved backward and sharpened to a point. They are surprisingly large (about 1 mm long) and fit snugly into the epigynum of the female, so that the tip through which the sperm exit sits right at the entrance to the female's sperm storage receptacle, her spermatheca. When the male withdraws his palp the embolus cap almost always breaks off and remains in the female's insemination duct. This has the effect of impeding access by subsequent males, a tactic that we shall return to when we examine the reproductive strategies of males.

Because spiders have internal fertilization we have no difficulty understanding why males have an intromittent organ and females have an insemination duct. We expect these traits to be sex specific. However, discerning why male and female *A. aurantia* are so different in size, shape, and color is more difficult. Assuming that the trait combinations characteristic of each sex have evolved because they increase the Darwinian fitness of their bearers, it must benefit males to be smaller, less brightly colored, and relatively longer-legged than females. Similarly, females must benefit by being larger, more conspicuously colored, and relatively shorter-legged than males. This is a very different pattern of sexual dimorphism than we saw in elephant seals, great bustards, and our shell-carrying cichlids, and it tells us that yellow garden spiders must maximize their reproductive success using strategies that we have not seen in these other species.

To discover what these different strategies may be, we return to the large, conspicuous female I nearly blundered into that September morning as she sat splendidly in the middle of her orb web. At this time of year, early fall in southern Canada, it is likely that she was a mature female and that her impressively large and round abdomen was bulging with a clutch of maturing eggs.[8] In this region female garden spiders mature in late summer and survive, at best, only until the first hard frost, which typically arrives before the end of October. As the days grow shorter and colder, they produce successive clutches of 300–500 eggs, which they deposit on carefully prepared beds of silk suspended from the vegetation near their webs. Once the clutch is laid, the mother wraps it in layers of tough, papery silk, to form a large, spherical cocoon or egg sac about 2.5 cm in diameter. As the autumn progresses the eggs hatch inside their silk cocoons, and the tiny spiderlings grow until they undergo their first molt. Then, as winter closes in, they stop their development and enter an arrested metabolic state called *dormancy*. This is analogous to hibernation in mammals such as bats and ground squirrels in that it enables animals to survive through harsh seasons when food is not available. However, the huddled spiderlings are far from immune to destruction. During the long months between the time that the eggs are sealed into their cocoons in autumn and onset of warm weather in April or May, almost all of the cocoons are attacked and damaged by predators or parasites, and many disappear altogether, probably carried off by birds for food or nesting material. In all, not more than three in a hundred egg sacs survive the winter unscathed.[9]

Females preparing their egg sacs in autumn clearly face enormous odds, and the old adage, "don't put all of your eggs in one basket," surely applies. To reduce the odds of total failure females produce successive egg sacs as autumn progresses. In the laboratory, with warm temperatures and abundant food, they construct an average of four egg sacs before they die, and some manage as many as seven.[10] However, the number of clutches they can produce is constrained by the time and nutrients required to mature a batch of eggs. The eggs remain small and immature until the female has mated but then increase in volume ten- to twelvefold as she adds yolk to them in preparation for laying.[11] Even under the benign conditions of the laboratory, it takes two to four weeks to produce a first clutch, with similar intervals for successive clutches. In the wild as autumn progresses the days shorten and temperatures cool, metabolic processes slow, insect prey becomes increasingly scarce, and females are at constant risk of mortality. Given these conditions intervals between clutches are likely to be longer and the total number produced considerably less than in the laboratory.[12]

One might think that females in the wild would do better to mature earlier in the season so that they would have time to produce more batches of eggs. However, to do so would mean maturing at a smaller size, which itself carries a fecundity cost. Like most spiders and insects garden spiders do not grow after they molt to the adult stage, and so the size of their adult skeleton depends entirely on growth achieved during the juvenile stages. For any given individual maturing earlier would mean ceasing growth earlier and thus molting into a smaller adult. Because the number of eggs a female lays in each clutch increases as her body size increases, the gains she might accrue from earlier maturation would be at least partly nullified by a reduction in her potential fecundity.[13] Females thus have to balance the benefits of breeding early (the possibility of having more clutches) against the benefits of growing larger (laying more eggs per clutch). The fastest-growing females are able to mature early and at a large size, and so they reap both advantages. However, females with slower growth face a trade-off between maturing early at a smaller size or postponing maturation so that they can continue to grow. Most opt for continued growth and so mature later than their faster-growing peers, which suggests that the advantages of larger size generally outweigh the advantages of earlier maturation.[14]

Most females mate either during or very shortly after their final molt and well before they have eggs ready to be laid. (The male's sperm remain inactive in the female's seminal receptacles until she is ready to *oviposit* several weeks later and can be stored for several months.) Mating occurs while the female hangs upside down in the center of her web, often in bright sunlight in the middle of the day. Although this is highly conspicuous behavior, it is seldom observed because the couplings are fleeting and rare: each copulation lasts only three to eight seconds, and most females mate with only one or two males during their lifetimes.[15] Once a female has mated and acquired sufficient sperm to fertilize her eggs, she is not usually receptive to male advances and is likely to treat approaching suitors as prey rather than entertain courtship attempts.

The long lag required to produce a clutch after the first mating places a premium on mating as soon as possible after reaching maturity. To ensure that suitors will be on hand (or more accurately, on web) at this crucial time, female *A. aurantia* employ an effective "come hither" tactic as they approach their final molt. They first construct a special molting web that lacks the sticky capture spirals of normal webs and is surrounded by extensive, loose, barrier webs. Once this molting web is complete, the female stops feeding and hangs motionless from the web for several days as she prepares to shed her old exoskeleton. While waiting to molt, she emits airborne pheromones that attract wandering males from afar.[16] The tiny males scramble through the tangle of vegetation toward the female and eventually contact the silk barrier strands radiating from her web, which are also likely impregnated with pheromones (although this has not been demonstrated specifically for *A. aurantia*). After climbing eagerly until he reaches the orb itself, each male settles down to wait, often for several days, until she undergoes her final molt and is able to mate. As time passes expectant males accumulate near or on the web, and most females have two or three waiting suitors by the time they undergo their final molt (in our populations, some had as many as seven).[17]

In the lives of mature females matings are infrequent and brief, construction of egg sacs takes only a few hours, and actual egg laying (*oviposition*) takes less than ten minutes; so these tasks occupy relatively little time.[18] Most of a female's time is devoted to capturing the food she needs to nourish her eggs and to produce the large volume of silk required for each cocoon. Like other orb-weavers yellow garden spiders employ their large, orb webs as stationary nets for harvesting insects from the air. The aim is to capture large, active prey such as grasshoppers and bees while at the same time to conserve as much energy as possible for allocation to eggs and cocoons. The female's strategy is to remain as immobile as possible, relying on her web to catch the prey rather than being an active hunter herself, and to that end she spends most of her time hanging immobile in the center of her web, waiting for prey to blunder in.[19] Other than subduing prey once they are tangled in the sticky spirals of her web, the female's main daily task is web construction and maintenance. The effectiveness of the orb web as a capture net depends on the tension in the silk spokes and spirals and the stickiness of the capture spirals. Wind, rain, dew, and struggling prey damage and degrade the web, and so webs have to be repaired and rebuilt frequently, usually every day. Web reconstruction occurs mainly overnight under the cover of darkness and is a remarkably efficient process that the spider does by rote in only twenty to thirty minutes. She often reuses the basic suspension frame from her previous web, removing the spiral silk and damaged spokes by consuming the silk so that its proteins can be reused. If her capture success has been low she may change the placement of her web and begin anew, but she is unlikely to move more than a half meter from her original site.[20]

Most predators that use ambush tactics to capture prey rely on camouflage to remain undetected until their prey are close enough for capture. A rattlesnake coiled under a bush or a stonefish mimicking a rock on the ocean bottom are familiar examples. Among spiders, crab spiders are excellent examples of this. They sit motionless on flowers, fruit, leaves, or bark waiting to capture passing insects, and their colors and patterns often blend completely with the colors of the substrate on which they sit. Although garden spiders are also ambush predators, they are clearly not using the same tactic. All *Argiope* females are brightly colored and spend their time at the center of their webs where they seem very conspicuous. They also enhance the conspicuousness of their webs with highly reflective, thickened patches of silk called stabilimenta. (In *A. aurantia* the stabilimentum takes the form of a vertical stripe that crosses the hub, with circular patch at the hub itself, and juveniles and males have these as well.) We can speculate about why these spiders are so conspicuous, but to be honest, no one really knows. Thinking about great bustards and cichlid fishes, one might suppose that the bright coloration serves for mate attraction. However, orb-web spiders have poor color vision and little visual acuity. They communicate primarily by chemical signals (pheromones) and substrate-born vibrations and at close range, by touch and taste.[21] Bright coloration is therefore unlikely to be an effective sexual signal. A more likely explanation is that garden spiders make themselves and their webs conspicuous so that large insects and birds will see them and avoid flying into and damaging their webs.[22] This is especially important for the large, mature females because their webs are the largest and are placed highest in the vegetation. Mature males, on the other hand, have no need for conspicuousness because they do not construct webs at all. Their strategy for achieving reproductive success is entirely different from that of females and, as we shall see, depends on rapid maturation, small size, and inconspicuousness.

The large size and predatory nature of mature female *A. aurantia* have major consequences for their males, and, not surprisingly, the males have evolved tactics to reduce their risks and increase their chances of success in their interactions with females. Having located a female's web and clambered up to the orb (not insignificant obstacles), the tiny males adopt one of two tactics, depending on the condition of the female. If the female has not yet completed her final molt, the male will have to wait to make his mating attempt. This is not necessarily a bad thing. A male that arrives on the web of a female shortly before her final molt and finds himself to be her only suitor is truly in luck. If he waits patiently until she molts, his chances of successfully mating with her are very high. The newly molted female is pale and soft and hangs by a silk dragline beneath her shed exoskeleton while her cuticle expands and hardens. Her soft cuticle prevents her from effectively moving her legs or her fangs, and for fifteen to twenty minutes she hangs helplessly from the web. This is what gives the male his chance of success. His tactic is to dash in and insert his palps while the female cannot defend herself. Spider biologists call this *opportunistic mating* to distinguish it from mating with a fully mature female that is capable of attacking and even killing unwanted males.[23] Our fortunate male moves to the center of the web and drops a dragline down to reach the hanging female. Once he contacts her body he scrambles frantically back and forth over her, touching the ends of each of her legs, her pedipalps, and the margins of her abdomen. Eventually he orients himself on the underside of her abdomen, facing her, with his prosoma over her epigynum. He then rears up and reaches beneath himself to insert a palp into one of the female's paired genital openings. This is somewhat of a trial-and-error effort because he cannot see what he is doing, and his embolus has no sensory or tactile cells that would enable him to taste or feel his way to the appropriate position.[24] I imagine the process as somewhat akin to trying to fit a key into a lock on a dark night—a surprisingly frustrating experience. If he eventually succeeds in inserting his palp (and most do), he quickly transfers his sperm (typically in less than five seconds) and withdraws his palp, almost always leaving the broken embolus tip behind.[25]

Now the job is half done. With one insertion the male has probably transferred only enough sperm to fertilize about three-quarters of the female's eggs. He has to insert his second palp to provide sufficient sperm to fertilize the rest. With the female still immobile the male quickly re-orients himself along her abdomen and inserts his second palp into her other insemination duct. This is his last living act. Within seconds after he has made his second insertion, he curls his legs under him and ceases to move. His heart rate slows, and he dies, hanging by his inflated palp from the female's epigynum. Once the female's cuticle has hardened sufficiently to allow her to move, she climbs back up her molting thread, resumes position at the hub of her web, and pulls her dead mate out from beneath her. In doing so she invariably tears him away from his swollen palp, and his embolus cap remains wedged in her insemination duct, blocking the entrance to her spermatheca. Although one might imagine that the females are somehow killing these males during copulation, this is truly not the case. The males always die spontaneously during their second insertion with no complicity by the female. Matthias even observed one male that mistakenly (one assumes) inserted his second palp in the carcass of a meal-worm larva that had been provided as food for the female. This deluded male curled up and died with his palp stuck in the dead mealworm.[26]

Spontaneous death during mating is certainly an odd tactic and not one that is typical of males in the animal kingdom. In most species males try to mate with as many females as possible. However, spontaneous male death seems to occur in all *Argiope* species, and males in many other species of orb-weaving spiders are complicit in their own deaths when they provoke or at least do not try to avoid cannibalistic attacks by their mates.[27] Why would males do such a thing? Some researchers have postulated that they may be sacrificing themselves to provide food for their mate so that she can produce more or better quality eggs. However, this seems an unlikely explanation because spontaneous male death and suicide by sexual can-nibalism occur only in species where males are much smaller than females and do not constitute much of a meal for her. In these species the male bodies simply do not contain enough energy or nutrition to detectably augment the female's fecundity or the quality of her eggs.

If the dead males provide little more than hors d'oeuvres for their mates, the explanation for male self-sacrifice must lie elsewhere. The key, it seems, is competition with other males. So far we have imagined our male garden spider alone on the female's web but unfortunately for him, this is unlikely to be the case. Males reach sexual maturity about two weeks earlier than females (table 6.1) and immediately set off in search of potential mates. In the wild they begin accumulating on the webs of females about a week before the latter are ready to undergo their final molt. Thus, by the time a female is available for mating, the waiting males usually find themselves in competition with other males. In our populations 92 percent of the waiting males faced at least one competitor, and about a quarter found themselves competing with three or more rivals. Under these circumstances even if a male is fortunate enough to mate with the newly molted female, he faces the prospect of having his efforts undone by competition of his sperm with the sperm of other males. Females can store the sperm from several males at once, but small as he is, our male spider has the ambition of fertilizing all of the hundreds of eggs that his mate will lay. His large bulbous palp with its long sclerotized embolus, one of the largest in all the orb-weavers, is ample evidence of this. However, he is playing a zero-sum game in competition with rival males: any fertilizations gained by other males are fertilizations lost to him. To prevent other males from inseminating his female, he tries to block her insemination ducts by leaving behind his embolus cap after his first insertion and by blocking the way with his entire body after the second. We have some evidence that this tactic works. If other males are present they try to pull the dead male away from the female's epigynum, but the first male is stuck firm and the path is blocked. The embolus cap itself does not prevent subsequent males from inserting their palps, but if a female has received only one insertion, subsequent suitors will avoid inserting in the duct that contains the embolus cap of the previous male, probably because the embolus cap reduces their fertilization success.[28] In species such as redback and black widow spiders, where females actually kill and eat their mates during copulation, males that are killed and consumed fertilize more eggs than those that escape, probably because they copulate longer with the female. All of this suggests that self-sacrifice by male spiders is primarily a strategy to enhance fertilization success in the face of potential sperm competition from other males.

Male yellow garden spiders much prefer opportunistic matings to the riskier alternative of attempting to court and mate with a fully mature female.[29] To have any hope of soliciting a mating with a mature female rather than a predatory attack, the hapless male has to tread carefully along the dry silk spokes of the female's web. He approaches the hub gradually and stops repeatedly to send vibratory signals that he is a suitor rather than a potential meal. The web is usually built at a slight angle to the vertical, and the female sits at the hub on the underside of the web while the male approaches carefully along its upper surface. The female may vibrate the web aggressively to deter the male and may even attack him (this happens about 27 percent of the time), but in spite of these deterrents, almost all males manage to reach the hub intact. Once at the hub the male drops down to the female on a silk security thread and begins a truly frenetic courtship. When he contacts the female, he scampers rapidly over her, tapping the ends of her legs repeatedly with his front two pairs of legs and running again and again over her body to do so. He taps her abdomen and her pedipalps and gradually seems to focus his attention on her front legs. If the female has allowed him to get this far, she generally raises the front of her body so that the male can climb onto her ventral surface. While clinging to her with his front legs, he probes with his palp until he gains an insertion, holds on for the several seconds it takes to ejaculate, and then attempts to jump free. This is the dicey bit. More than three quarters of the males are attacked during this insertion and a third of these are killed before they can escape. Typically, the female collapses over the inserted male and rapidly wraps him in silk while he is still inserted. The ability to ejaculate quickly and jump free with alacrity is clearly at a premium.

The males that manage to avoid a silky death on first insertion climb back up their security thread to court a second time. About a quarter of these males are attacked, and others make several unsuccessful attempts to insert before giving up and jumping away. Their behavior looks like blind incompetence fueled by fear, but we really have no idea why these males fail. In the end only about a third of males manage to achieve their second insertion. Just as in opportunistic matings these males then die spontaneously while *in copula*. Although this self-sacrifice works as a mate guarding tactic when the female is newly molted and still soft, it seems pointless here because mature females pull the dead males out within a few seconds (our average was eight seconds), and they are very unlikely to mate again in any case. So why do the males not jump off and drop safely away as they do after their first insertion? In several other *Argiope* species death during copulation appears to benefit the male by increasing the duration of copulation and the number of sperm he is able to transfer,[30] but we could find no evidence that the number of sperm transferred or fertilization success increased with copulation duration in *A. aurantia.* An alternative explanation for the spontaneous demise of our males is that it has evolved because it benefits them in opportunistic matings and simply carries over to matings with mature females. Spontaneous death can persist in the latter context because, although males may not gain anything by dying, they also have nothing to lose. Having achieved two insertions and lost both emboli, they cannot mate again whether they survive or not. Thus, strange as it may seem from a human perspective, the male spiders have nothing to gain by further survival, and so their death on second insertion in no way reduces their Darwinian fitness.

The life of a male yellow garden spider is clearly fraught with daunting challenges. If he hopes to father any spiderlings he will have to contend with rival males, cannibalistic females (literally "femmes fatales"), or both, and his personal reward for success is death *in copula.* As if that were not discouraging enough, his biggest obstacle is actually finding a potential mate in the maze of herbage that is his habitat. As we see below, the chances are good that he will get eaten or lost or simply starve along the way. It seems likely that being small, relatively long-legged, and inconspicuous helps him to overcome these obstacles and win a mate. Otherwise, why would male garden spiders not be similar to females in size and appearance, or even larger and more flamboyant than females, like male bustards, elephant seals, or our shell-carrying cichlids? The answer to this question has proven elusive, and I admit that the puzzle is not yet completely solved. Nevertheless, we have discovered some of the benefits of the typical male morphology in garden spiders, and we can make sensible inferences about several others.

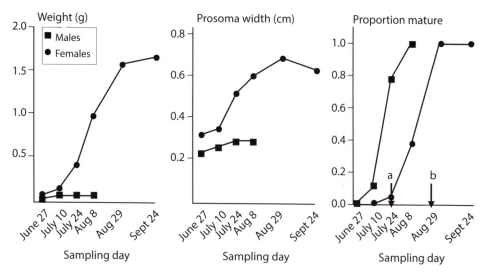

FIGURE 6.2. Mean weight (left panel), width of the prosoma (center panel), and proportion mature for male and female *Argiope aurantia* captured in an old field near Aiken, South Carolina. Note that no males were found after August 8. Arrows indicate (a) the date that males were first observed on female webs and (b) the date that egg sacs were first observed. Data are from Howell and Ellender (1984).

The one clear advantage of being small is that it enables males to stop growing and become sexually mature earlier than females. Spending less time as juveniles increases the probability that the males will survive to sexual maturity, and maturing before females increases the likelihood of obtaining an opportunistic mating. To that end males go through fewer molts, grow much more slowly, and reach maturity much earlier than females (table 6.1, figure 6.2). By the time the spiderlings emerge from their egg sac in spring, the females are already larger than males, and they continue to grow about 30 percent faster than their brothers until the males mature and stop growing altogether. The benefits of higher survival to maturity and earlier maturation could be sufficient to explain the small size of male *A. aurantia* relative to their mates, but it seems likely that small size and relatively long legs contribute to their fitness in other ways as well. After their final molt the life styles of males

diverge radically and irrevocably from those of their sisters. The newly
molted males abandon their lives as web-builders and begin wandering
through the vegetation in search of a mate. This behavior, called roving,
is unique to mature males. Without webs, the males cannot capture prey,
and so they do not eat again unless they can scavenge food from the web
of a potential mate. The longer they search for a mate the thinner they
get, and their abdomens visibly shrink as their search continues. This is
when being small, inconspicuously colored, and relatively long-legged is
likely to be most advantageous. The roving males tend to be active during
the day when they are vulnerable to visually-hunting predators such as
wolf spiders, birds, and lizards (although no lizards come as far north as
our study sites in Quebec).[31] When on the ground they move slowly and
carefully, stopping often, and one can easily imagine that small size and
muted coloration would make them less conspicuous to predators. Being
small and relatively long-legged also enhances the speed and efficiency
with which they can climb the stems of grasses, herbs and small bushes.[32]
Adult females are far too large to climb such stems efficiently and seldom
move from their web sites, but the males are good climbers and readily
scamper up. Small size and relatively long legs also enable males to take
advantage of a mode of locomotion that spider biologists call bridging.
To make a bridge the spider climbs to a position on the plant that enables
it to catch the breeze. Once there, it lets out a strand of strong, flexible silk
which is wafted on the breeze as it lengthens. Eventually, the floating end
of the silk catches on a neighboring plant. The spider then pulls the line
taut, forming a single-stranded bridge between the two plants. To move
across, the spider hangs by its legs from the line and runs across upside
down. Bridging is a highly effective technique because it allows spiders
to move from plant to plant without climbing down, walking across the
ground, and then climbing back up again. Small spiders with relatively
long legs (especially front legs) are the best bridgers, and spiders that
weigh more than 0.1 g rarely bridge at all.[33] Female yellow garden spiders
exceed this threshold quite early in their development, and adult females
never bridge. In contrast, males use bridging as a primary mode of movement during their roving phase. The advantages of bridging and climbing

clearly favor males with much smaller bodies and proportionally longer legs than females, and it seems likely that selection for these attributes contributes to sexual dimorphism in *A. aurantia*.[34]

A slightly different advantage of small size is that smaller animals require less food. Roving males do not eat once they have abandoned their webs and must fuel their mate searching with the energy they acquired prior to their final molt. They are akin to motorists trying to drive as far as possible on one tank of gas: it is best to have a small car with a big gas tank and to fill the tank as full as possible before you start. However, because smaller animals have higher metabolic rates, which is analogous to having a higher idling speed in a car engine, the advantage of small size only kicks in if smaller animals store proportionally more energy (i.e., have proportionally larger gas tanks) or if they are able to use their stored energy more efficiently during mate search.[35] We do not have good information about proportional energy storage in yellow garden spiders, but the little we have suggests that the proportion is relatively constant across the range of adult male sizes.[36] In other words smaller males do not seem to have proportionally larger gas tanks than larger males. However, the efficiency of moving through the vegetation does depend on size, as we saw above, and so it seems likely that having small bodies with relatively long legs reduces the energetic costs of roving and so enables males to rove longer and cover larger distances in their search for mates than they could if they were larger.

We have now accumulated quite a good list of reasons why male garden spiders might benefit by being much smaller than females, relatively long-legged, and cryptic, at least until they actually find a potential mate. However, once a male reaches a female's web, the situation changes dramatically. If the female is not yet mature the male is likely to have to compete with other waiting males to mate with her during the short window following her molt. If the female is already mature he is less likely to face competition from other males, but he will have to entice the female to mate and avoid her attacks while doing so. In both cases only a subset of males succeeds in mating, and so there is ample opportunity for sexual selection to act. The examples from previous chapters would lead us to expect that sexual selection through mate choice or male-male interactions would align with the patterns of sexual dimorphism. In other words males that have small bodies, relatively long legs, and more cryptic coloration should have higher mating success under one or both situations. Surprisingly, this does not seem to be the case. No one has looked for sexual selection on male coloration in garden spiders, but their poor visual acuity, lack of color vision, and absence of visual signaling argue against color being important for mate choice or male-male competition.[21] Size does influence male mating success in opportunistic matings—but not in the expected direction: the successful males tend to have both larger prosomas and longer legs than their rivals and so are larger overall.[37] Males with longer legs are more likely to mate successfully and less likely to lose a leg in struggles with other males, but this seems to be because they are larger overall, not because they have long legs relative to their body size. Although the competing males typically cohabit on the female's molting web for several days, the advantage of larger size seems to accrue only when the female actually molts. Waiting for this is like a game of musical chairs for the males. They move around the web, with no male consistently retaining a position nearest the female. Occasionally they fight with each other, but fights are rare until the female begins to shed her skin. Then, as though the music has stopped, all the males rush in and fight vigorously to gain access to her epigynum. The size advantage occurs during these final struggles, when larger males successfully jostle aside their smaller rivals.[38]

When attempting to mate with mature females males face a different set of challenges, but again, there is little to indicate an advantage for small, relatively long-legged males.[29] As far as we can discern, neither the male's body size nor the length of his legs influences his chances of being killed by the female. These traits are also unrelated to his success in obtaining insertions or the number of eggs he fertilizes after one or two insertions. In sum, it seems that mature females are indifferent to the size of the male that fathers her offspring. If they do discriminate among suitors it must be on the basis of chemical cues or behavior rather than body or leg size. Male size matters only when females are mated opportunistically and males compete with each other. Since this boils down to physical contests between males, it is not surprising that larger males are more likely to be successful.

To reconcile the lack of congruence between sexual selection during mating interactions and the pattern of sexual dimorphism we need to see mating interactions in a broader context.[39] In highly dimorphic orb-web spiders the biggest challenge faced by males is finding a female of the appropriate age and disposition to mate. Our best estimates suggest that 75–90 percent of males fail in this search.[31] The males that actually arrive at a female's web are thus only a small subset of those that started out. Any sexual selection that occurs during mating interactions on the female's web can act only on males that survived the roving phase. Males whose morphology, physiology, or behavior put them at increased risk of mortality while roving or make it less likely that they can locate a female's web, are unlikely to be among the males competing on the female's web. If selection during roving favors smaller males with relatively longer legs, as seems to be the case, the males that reach female webs will tend to be small males with long legs, even if some other combination of traits would serve them better in the mating arena. The advantage of larger size in male contests on the web no doubt contributes to setting the optimal size for yellow garden spider males by disfavoring the smaller of these males. However, it does not cause male size to increase because larger size is disfavored by the selection that occurs before males meet their potential mates. For our male spiders this would include selection favoring avoidance of predation, efficient mobility, long-term survival without eating, and a keen ability to detect and follow female pheromone cues. Given the clear advantages of opportunistic matings for males, we might also add early maturation to this list because this increases the chances that males will find females before their final molt.

We have now come full circle, back to the large, conspicuous female garden spider hanging at the center of her web on that sparkling September morning. In all likelihood she had already mated with at least one male and was now yolking up her first batch of eggs. No males were in evidence and most had probably already died in a futile search for a mate or, for a lucky few, in a final copulatory embrace. In the context of our own experiences as humans, and compared to the three species described in previous chapters, these male and female sex roles seem decidedly odd. Male garden spiders do not defend harems of females as we saw in elephant seals and shell-carrying cichlids. Nor do they follow the great bustard strategy of executing flamboyant displays to attract females to come to them. In these species successful males are likely to have many mates in a given mating season, whereas females mate with only one or a few males during the same period. In marked contrast to this, dwarf male garden spiders accumulate on the webs of giant females and compete with each other to garner a single mate. Most perplexing of all is that these males die spontaneously while mating. Their ultimate goal is to mate with one and only one female.

This improbable mating strategy, where females mate with more than one male but males mate with only a single female is common in spiders, and as we shall see in subsequent chapters, it is a general characteristic of animals in which males are dwarfs relative to their females.[40] In these species females tend to be widely dispersed, rare, and difficult for males to locate. Males mature early and spend a long time, often much of their lives, searching for a mate, and they run considerable risk of failing in this search. Most fall prey to predators during their search, while others may simply fail to locate a mate before running out of energy or otherwise suffering from inevitable physiological damage and decline. The elusiveness of females means that once a male manages to locate one potential mate, he is unlikely to find a second. Given that his probability of future reproductive success is negligible, it behooves him to invest everything he has in his current mate, particularly if this reduces the chances that his efforts will be undone by subsequent suitors.[41] This is why yellow garden spider males come equipped for only two copulations and try to execute both with the first receptive female that they find. They maximize their Darwinian fitness by dedicating all of their efforts to fertilizing as many of her eggs as they possibly can because their chances of successfully locating and mating with a second female are vanishingly small. Spontaneous death during copulation is a rather extreme means of achieving this objective, and it is rare even among spiders. However, it is typical for dwarf males to dedicate their lives to a single female once they have found her. They may be killed and eaten by their mate or die spontaneously shortly after mating, as in many spiders, but more commonly, they simply abandon free-living and set up house on or in their mate, devoting the rest of their lives to doing nothing but providing her with sperm. Spiders are the only free-living, terrestrial animals to have adopted this strange and extreme allocation of sex roles. However male dwarfism (or female gigantism[42]) with its associated sexual divergence in age at maturation is by no means unique to spiders. The same pattern of extreme sexual dimorphism can be found in at least twenty-two other animal classes distributed across twelve phyla,[43] and we explore the lives of several of these extraordinary species in the chapters to follow.

CHAPTER 7

Blanket Octopuses

Drifting Females and Dwarf Males

Spider biologists argue, with considerable justification, that the extreme sexual dimorphism found in many species of orb-web spiders is the result of females becoming giants rather than males becoming dwarfs.[1] However, in numerous groups of marine animals the epithet "dwarf male" is clearly appropriate.[2] In some of these species the females may be larger than one might expect when compared to closely related species, but the real standouts are the males. Not only are these fellows tiny when compared to their mates, they are also stripped-down versions of fully functioning, independent adults. In essence these dwarf males have become ultraspecialized, sperm-delivering units. They generally bear scant resemblance to their larger, free-living relatives or to their mates. In these cases it is obvious that the males must have become smaller and more structurally reduced over evolutionary time. The term "dwarf" refers only to their small size, but, as we shall see, this diminution tends to be accompanied by similarly radical reductions in structure and function. This and the following three chapters describe four very different kinds of marine animals that have independently adopted this extreme division of sexual roles. Because species with dwarf males tend to occur in low densities and in generally inaccessible habitats, such as the open ocean or the ocean depths, they have been less intensively studied than the animals I have described in previous chapters. This means that each of these examples is necessarily somewhat less detailed and more speculative than those that have gone before. Nevertheless, these rare and extraordinary species serve as compelling examples of the most extreme sexual differences in the animal kingdom.

The first example is an octopus that makes its living cruising in the warm surface waters of subtropical and tropical seas around the globe. Its common name, blanket octopus, refers to the broad, translucent web that stretches between the long dorsal arms of the females (plate 15). This name is shared by four species in the genus *Tremoctopus*, three of which are so similar that they are distinguished primarily by where they occur: *violaceus* in the Atlantic Ocean and Mediterranean Sea, *gracilus* in the Pacific and Indian Oceans, and *robsoni* in New Zealand waters. Zoologists can also distinguish them by counts of gill filaments and pairs of suckers on their copulatory organ, but these are very subtle distinctions. Until recently all three were considered to be *Tremoctopus violaceus*. The fourth species, *gelatus*, is noticeably different. Its name refers to the composition of its tissues, which are gelatinous and quite transparent rather than firm and muscular as in the other species. It also differs in color, shape, and size from the *violaceus* group, so is easily distinguished. Little is known about *gelatus* because it is rarely found, but it seems to occur in the tropical and subtropical zones of the Atlantic, Pacific, and Indian Oceans. Because we know so little about *gelatus*, my descriptions in this chapter refer only to the better known *violaceus* group.[3]

Octopuses are odd animals by almost any standard, and blanket octopuses are no exception. They are so different from any of the animals that I have described so far, and indeed from any of the animals that most of us ever think about, that we should probably take a moment to consider what an octopus is.[4] Octopuses are in the taxonomic order Octopoda within the phylum Mollusca, the phylum that also includes shelled animals such as clams, oysters, limpets, and snails. Together with squid, cuttlefish, and nautiloids, the octopods comprise the class Cephalopoda, so called because they have arms that attach directly to their heads (cephalo = head, pod = foot or appendage). The thick, muscular foot that anchors clams, oysters, and abalones to the substrate (and that we humans find so tasty) has been modified in cephalopods to form a cluster of highly mobile appendages. Octopuses, as their name suggests, have eight of these. The arms have rows of suckers that enable octopuses to hold and manipulate prey and also to attach themselves to rocks, pilings, and even the glass sides of aquariums. The circle formed by the bases of the arms surrounds the octopus's mouth with its prominent, vicious-looking beak and tongue-like, tooth-covered radula.[5] Together the beak and radula enable the octopus to bite and rip its prey to bits. Behind the mouth lies the head supporting a pair of enormous eyes, followed by a narrower neck region, and then a sac-like body containing the reproductive and digestive organs. The body is encased in a muscular envelope called the mantle that is generally cone-shaped and often forms a distinct collar around the neck region.

One of the characteristics that makes octopuses and other cephalopods so fascinating is that their large eyes are remarkably similar to our own. They have a cornea, lens, diaphragm, and retina and are capable of forming an image and focusing. The structure of the retina, the type of visual receptor cell, and the mechanism of focusing are all different from ours, but it is clear that both types of eyes operate using the same optical principles and probably produce a similar view of the world. Octopuses also have large, complex, and highly differentiated brains. The weight of their brains relative to their bodies matches that of fishes and reptiles, and at least some species are capable of quite complex learning. For example, in captivity, the common octopus has displayed a remarkable ability to learn and remember tasks such as attacking, avoiding, or eating objects based on color, shape, texture, or size. They can open jars, play with toys, and even recognize their keepers.[6] Who would have thought this of a mollusk? Clearly one should be careful not to judge an animal by the phylum it keeps.

Most octopuses are active hunters that hide by day and come out at night to seek their prey in the nooks and crannies of the ocean bottom. They use their arms for moving over the bottom or paddling slowly through the water, and when they need to move quickly, they propel themselves by expelling a jet of water through a muscular funnel, another derivative of the molluscan foot. The vast majority prefer shallow coastal areas, or at most, the somewhat deeper expanses of the continental shelves. A rare few are truly abyssal and roam the deep ocean bottoms at depths of up to 7,000 m (4.35 miles). Unlike their relatives the squid, which are active swimmers and often hunt in large schools, octopuses are usually solitary and clearly prefer to have their feet (or rather, arms) on the ground. Blanket octopuses are one of the few exceptions to this rule. They belong to an unusual clade of octopuses called the Argonautoidea, named after the argonauts of Greek mythology[7] because of their habit of drifting in the ocean currents, hundreds of kilometers from land. All of the Argonautoidea[8] have dwarf males, and in all but one species these are less than 5 percent the length of their mates when mature. Most cephalopods have few externally obvious sexual differences, so this extreme sexual size dimorphism is highly unusual. Blanket octopuses have the most extreme sexual size dimorphism of all, with males only about 2 percent as long as females: the tiny males are less than 4 cm long and weigh about a quarter of a gram, whereas the females reach lengths of up to 2 m. The entire mature male is about the size of one the female's eyes.

Male blanket octopuses are true dwarfs. In addition to being minuscule relative to their females, they are among the smallest of all octopuses. Two species in the genus *Octopus* are as small or smaller (both sexes), and two closer relatives in the genus *Argonautus* have equally small males, but among the 200 or so species in the Octopoda, such tiny size is very rare.[9] In marked contrast to their males, female blanket octopuses are among the largest of all octopuses, and so we might equally consider them to be giants. Only two species, the Pacific giant octopus and the seven-arm octopus, another Argonautoid species, are known to be larger. The net result of combining giant females with dwarf males is that mature female blanket octopuses weigh 10,000 to 40,000 times more than their mates.[10] Such a difference between sexes beggars the adjective "extraordinary" and, as you might expect, is associated with huge differences in the life histories of males and females.

All blanket octopuses start out as small, yolky eggs that hatch into miniature versions of their adult selves, albeit with stubby little arms. The two sexes can be easily distinguished as soon as they hatch because the tiny males seem to lack one of their arms. This is because, rather than developing into a typical octopus arm, the third right arm of males is coiled inside a pouch between the bases of the second and fourth arms. As the tiny males grow, this pouch enlarges until it bulges out between the funnel and right eye, giving males a distinct, asymmetrical appearance (plate 16). This modified arm, called the *hectocotylus*, will serve as a copulatory organ for depositing sperm inside the female's mantle cavity if the male is fortunate enough to survive to maturity and find a mate. The juvenile octopuses grow rapidly while drifting in the surface waters, apparently feeding on other small animals that they encounter by chance. Their primary diet seems to be sea slugs and sea snails supplemented with the occasional small fish. They are weak swimmers and, when small, rest from time to time by clinging to the bells of drifting jellies. Perhaps associated with this activity they pick up pieces of the tentacles of the highly venomous Portuguese and Pacific man o' war jellies. Small blanket octopuses, up to lengths of about 7 cm, almost always hold pieces of man o' war tentacles in the suckers of their four dorsal arms. When threatened, they curl their arms back over their heads exposing the battery of these tentacle fragments (this is what the male in plate 16 is doing). The stinging nematocysts in these fragments discharge when the octopuses are handled, which suggests that they make effective weapons both for defense against predators and for subduing prey.[11] Males use this tactic throughout their lives, but females quickly outgrow it. Their arms lengthen, and their web expands disproportionally as they grow until it spreads like a cape between their outstretched dorsal arms.[12] Although females can swim actively by undulating their arms and jetting water through their muscular funnel, it seems that much of their life is spent floating rather languorously in the sunlit surface waters of the open ocean (called the *epipelagic zone*), passively fishing for small invertebrates among the drifting plankton.

Male and female blanket octopuses part ways quite early in their developmental trajectories.[13] Males never develop much of a web and retain juvenile dimensions throughout their lives with relatively short arms, big heads, and short, broad mantles. As they grow, their hectocotylus grows disproportionally in its pouch until it is longer than the male's body and

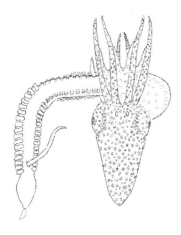

FIGURE 7.1. A male blanket octopus (*Tremoctopus violaceus*) with his hectocotylus extended (length of body plus hectocotylus = 13.8 mm). The pouch in which the hectocotylus developed can be seen as the bulge beneath his right eye. Reprinted by permission from Thomas (1977), figure 6d.

other arms combined (figure 7.1). At the same time the male's single, large testis starts producing sperm that he packs into a tough, elastic capsule called a spermatophore. When finished the spermatophore contains all of the male's sperm plus a coiled ejaculatory apparatus that will propel the sperm into the female's oviduct when the spermatophore ruptures. The males store their spermatophore in a special sac, called Needham's sac, until they meet with a receptive female. Only then do they pass it to their hectocotylus for transport to the female's mantle cavity. Sexual maturity is a gradual process for male blanket octopuses, but biologists consider males fully mature when they have their spermatophore prepared and secure within Needham's sac. Once that has been accomplished, the male's only goal in life is to find a receptive female.

While males are devoting their energy to growing their hectocotylyzed arms and making their spermatophores, females are putting their energy primarily into growing larger and larger. They do put energy into reproductive tissue, filling their single ovary with 100,000–300,000 immature eggs, but most of their growth is nonreproductive (*somatic*) tissue. We do not know how long females spend in this growth phase before maturation, but octopuses in general have relatively short lives for their size. Most apparently live only one or two years, and even the huge Pacific giant octopus, which may weigh more than 270 kg (close to 600 pounds), lives only three to five years in captivity.[14] Based on their size it is likely that female blanket octopuses continue growing for at least a year and perhaps for two or three years before becoming sexually mature. Their transition to maturity is marked by rapid yolking of the eggs and a concomitant increase in the size of the eggs until each is 2–5 mm long (the length varies among species and populations) and about 2 mm in diameter. The eggs mature in batches of 10,000 to 30,000, so that a mature female typically has a series of egg cohorts maturing sequentially. Spawning has not been observed in blanket octopuses, but based on the number of cohorts of eggs contained in captured animals and their relative stages of development, researchers surmise that the females probably spawn as often as once per day over a period of two to four weeks. The tactic of female blanket octopuses is to make many very small but very yolky eggs so that the embryos, although tiny, have ample food in their yolk sacs to fuel rapid early growth.

The mother protects her eggs from predation and prevents them from sinking to the cold depths by attaching them to her web as she spawns. To accomplish this she first secretes a calcareous substance that forms a long, hard, sausage-shaped rod attached to her web at the base of her first arms. The eggs are attached to this rod as they are spawned via long strings to which the egg stalks are joined. Each cohort then hangs suspended in multibranched clusters, like tiny grapes on a vine, until the embryos hatch and drift away.[15]

Spawning marks the final stage of the female's life. It is hard to fathom how this can happen to such a large and seemingly intelligent animal, but once the female's eggs begin to mature, her body begins a progressive physiological decline that is inevitably fatal. She senesces throughout the brooding period and dies shortly after the departure of her youngest hatchlings. This is typical of almost all octopuses. Females spawn their eggs, brood them in one way or another, and then die shortly after the eggs have hatched. We do not know how long blanket octopuses brood their eggs, but most octopuses brood for months, even years. The senescent phase of a female's life is thus likely to be quite prolonged, comprising up to a quarter of her lifespan.[16] While brooding, female octopuses do not feed, and most lose 50 percent to 70 percent of their body weight before finally expiring. They depend on the nutrients they have stored up before spawning to fuel their body processes long enough to see their eggs through to hatching. However, unlike vertebrates (recall our elephant seals, bustards, and cichlids), octopuses cannot accumulate fat reserves that can be drawn down during fasting. Instead they have to get the energy they need by breaking down their muscle tissue, essentially digesting themselves in the process. This may be one reason why female blanket octopuses grow so large and wait so long to spawn: they must accumulate sufficient biomass in their muscle tissues to provide the energy and nutrients required to add yolk to hundreds of thousands of eggs as well as support themselves long enough to successfully brood those eggs through to hatching.[17]

Large size probably also has other advantages for female blanket octopuses. It is likely that fecundity scales steeply with body size, as it does for most animals that lay large batches of eggs synchronously, and so larger females probably produce more eggs.[10] It may also be that female foraging efficiency increases with size and especially with web size. As females grow, the area of the unfurled web scales up as a function of the square of arm length rather than linearly, and so as females increase in length, the size of their webs increases disproportionally. If females obtain their food by sensing prey items that touch their web and mantle and simply sweeping them up with their fourth arms, as has been surmised,[18] than a larger web would likely mean greater foraging success. Growth would then benefit females because foraging success would increase disproportionately with body length. This is pure speculation, of course, because no one has actually observed females foraging, let alone measured their feeding rates as a function of web size.

Male blanket octopuses never grow extensive webs and apparently rely on their stolen man o' war tentacles to capture and subdue prey. Unlike females they continue to feed after they reach sexual maturity. Armed and ready, with their charged spermatophore in its Needham's sac and their enormous hectocotylized arm packed securely under their eye, they drift along waiting to encounter a female. We do not know if such encounters are left to chance or if the males actively search for females. Males have very large eyes, so it may be that they find females at least partially by sight. If so, detection must occur only at short range, relative to scale of the open ocean. Perhaps blanket octopus females, like female garden spiders, actively attract males by releasing pheromones. However, given the limited power of the tiny males to actively propel themselves over long distances, especially against winds or currents, such chemical cues would probably be effective only at relatively short range. It seems likely that the sexes rely mainly on time and currents to bring them close enough together for either chemical or visual cues to be effective. Such a strategy requires long lifespans after maturity for the tiny males. To manage this they have evolved a life history strategy that is, in some ways, an exaggerated version of what we saw in our male yellow garden spiders. They become sexually mature at a very small size and young age (probably a year or more before their sisters), spend a significant portion of their adult lives searching for a mate, and then die during mating. Unlike our spiders, however, male blanket octopuses can combine mate searching with foraging and so do not have to give up one for the other. Being able to eat as adults enables them to mature even earlier and spend relatively more of their lives searching for mates than male spiders can. Being small may also benefit them directly simply because they require less food than larger males and so can spend less time feeding and more time searching for mates than they could if they were larger.[19]

Male blanket octopuses also share with spiders the tactic of co-opting a secondary appendage for use as a copulatory organ. Even more than in spiders the morphology of male blanket octopuses is dominated by the enormous size of this copulatory organ, the hectocotylus, a modified third right arm (figure 7.1). When the male finds a receptive female his hectocotylus uncurls and emerges from its security pouch with the single spermatophore secured at its tip, poised to launch sperm into the female's oviduct. No one has ever observed the emergence of a hectocotylus or any other aspect of the mating behavior of blanket octopuses, so we have to piece the story together from the broken bits left behind. The broken bits in this instance are intact male hectocotyli found in the mantle cavities of females. These clearly tell us that the hectocotylus breaks off during mating. This happens in several other octopus species as well, and in some of these the males are able to regenerate hectocotyli within six or eight weeks. However, male blanket octopuses have never been found with regenerating hectocotyli. Nor have males ever been found without intact hectocotyli. This evidence, although only circumstantial, indicates that the tiny males must die during or shortly after their single mating. The hectocotyli, however, have lives of their own. They are so large and peculiar that the august French zoologist, Georges Cuvier, identified them in 1829 as some form of parasitic worm and gave them the scientific genus name of *Hectocotylus*.[20] Subsequent zoologists eventually realized the true nature of these supposed worms and speculated that they were released by the male in the water and sought the female out on their own, since the remaining parts of the male were never found with females. More detailed morphological studies have debunked this notion, and the prevailing opinion now is that the male somehow manages to deposit his hectocotylus on the female before it breaks away. A prominent expert on cephalopods, Professor Kir N. Nesis, opined poetically that the hectocotylus "breaks off at mating and creeps in, wriggling snake-like, through the female funnel into her mantle cavity."[21] Although this may well be true, no one really knows how the hectocotylus and its spermatophore get inside the female.

We do know that the hectocotylus remains inside the female long after the spermatophore has ruptured and the sperm have entered the female's oviduct because females are often found with one or more spent hectocotyli inside their mantle cavities. The presence of these reveals another quirk of the reproductive strategies of blanket octopuses. Females have adopted an unusual tactic to ensure that they have sufficient sperm to fertilize their enormous numbers of eggs in spite of the presumed rarity of mating encounters. Although females are not considered to be sexually mature until they contain mature, fully yolked eggs and are ready to spawn, they are clearly receptive to mating long before they reach this stage. Like most octopuses blanket octopus females have specialized glandular expansions of their oviducts that serve as sperm storage reservoirs, and they are apparently able to store viable sperm for months prior to maturing their eggs.[22] They are thus ready and able to capitalize on encounters with males during their long, prereproductive growth phase by mating when the chance arises and saving the sperm for later use when they are ready to spawn.

The lives of male and female blanket octopuses could hardly be more disparate. Odd they certainly are, but each sex is odd in its own way. The tiny males arm themselves with stolen man o' war tentacles, tuck away copulatory organs longer than their bodies, and drift through the ocean seeking a mate, only to die as soon as they achieve their goal. The giant and supremely graceful females continue to live and grow for many months or even years longer than males, spawn hundreds of thousands of eggs, brood them faithfully, and then die shortly after their eggs have hatched. Beyond the earliest juvenile stages the two sexes are so different that the only way they could be identified as the same species is through identification of the male hectocotyli left inside the mantle cavities of the females. To survive in the vast, open ocean that is their home, females have become large and long-lived, and males have become small and short-lived. Like our orb-web spiders the giant octopus females may mate with several males, but the tiny males die after a single mating. The parallels between yellow garden spiders and blanket octopuses are many: tiny males that mature early and have short lives; giant females that mature later and have longer lives; males that mate with only one female and then die; females that have very high fecundity and frequently mate with several males. Females of both species also mate and store sperm before they yolk up their eggs rather than waiting until they are ready to oviposit. This is a life history built for habitats where juvenile survival rates are poor, adult densities are low, and mature females are rare, dispersed, and difficult to find. Low offspring survival favors high fecundity and hence large size in females, and the scarcity of females favors males that mature early and put their time and energy into searching for mates and providing them with abundant sperm rather than growing large themselves. This is not the only way to cope with life in these sorts of habitats (after all, elephant seals have adapted to life in the open ocean via a very different strategy), but it is one that has evolved repeatedly in many different animal lineages. In the chapters that follow we will see this basic pattern taken to even more extreme forms in species where males have become so small and specialized for mate location that they forgo sexual maturity entirely until they find a mate. Only once they find a willing female do they begin to produce sperm, and to do so they have to become tenants on or inside the bodies of their mates.

CHAPTER 8

Giant Seadevils

Fearsome Females and Parasitic Males

"Giant seadevil" is certainly not a name that inspires affection. One imagines a huge, fearsome animal of some sort, perhaps with a gaping mouth full of hideous teeth to invoke the devilish image. Such an image is not far from the truth. Seadevil is the common name for a diverse group of decidedly unattractive anglerfishes found in the depths of the open ocean. As their name suggests, anglerfishes are predators that capture prey by "angling" with a lure. In most species the lure is suspended on a stalk so that it dangles or wriggles close to its owner's mouth. When an unsuspecting fish or squid or shrimp approaches to investigate this tempting morsel, the anglerfish opens its huge mouth and, in a flash, sucks in the hapless victim and chomps down with rows of dagger-like teeth. Most other types of anglerfishes lurk on the seafloor in relatively shallow water often disguising themselves as lumps of seaweed or coral. However, the ancestors of modern seadevils long ago left the realm of the shallow ocean floor and moved into the cold, dark ocean depths, a bleak and biologically forbidding habitat that marine scientists call the *bathypelagic* zone. The 160 species of seadevils have become so highly specialized for life in the deep that zoologists classify them as a separate suborder of anglerfishes, the Ceratioidei. Descriptive accounts often refer to them simply as deepsea anglerfishes or ceratioid anglerfishes, but I prefer their much more evocative common name, seadevil.[1,2]

Seadevils are characterized by having dwarf males, which is why I have chosen them as an example of extraordinary differences between males and females. However, it is not the dwarf males that have garnered the moniker "devil." That credit goes to the much larger females. When first discovered, whether floating dead at the surface, washed up ignominiously onto a beach, or salvaged as flattened by-catch from a fishing trawl, these bizarre fishes must have truly seemed like devils from the deep. They are clothed in scaleless, black or dark brownish skin, often marked by knobbly spines or tubercles. Most have fantastically large heads (typically comprising more than 40 percent of their bodies) and unusually deep bodies, so that their overall profile has a decidedly rounded or oval look. Their huge mouths gape, revealing rows and rows of long, viciously pointed teeth of various sizes, clearly designed to seize large prey with a fatal grip. If these characteristics were not sufficient to make a superstitious ancient mariner pull back in horror, the soft, rather amorphous texture of the body tissues and the strange, fleshy appendage dangling from the snout would probably do the trick. This was no ordinary fish, and clearly not one to be trifled with.

In truth, in spite of their monstrous appearance seadevils pose no threat to humans. The bizarre morphology of the females and the tiny size of their mates are simply adaptations for life in the bathypelagic zone where, belying their name, seadevils lead rather dull lives. The bathypelagic zone extends down to 4,000 m (2.5 miles) below the ocean surface, and its upper bound is defined by the deepest penetration of sunlight, which varies from as little as 300 m (1,000 feet) below the surface to more than 1,000 m (3,300 feet) depending on latitude and water clarity. Without sunlight there can be no photosynthesis, so no organisms can convert carbon dioxide to organic matter at these depths. Without this primary productivity and with little input of detritus from above, food is scarce, and animal densities are necessarily low. Temperatures are numbingly cold (2°–6° C), and the pressure ranges from 30 to 400 times that at sea level. Through this cold, dark world the ponderous female seadevils move like ghosts. They are indifferent swimmers, with only a few tiny fins, poorly developed body musculature, and soft watery tissues. If pressed they can probably propel themselves a short distance by forcibly ejecting water through their gill openings, which have been modified into narrow tubes, perhaps for that purpose. However, it seems likely that they spend most of their time drifting lethargically, like anglers idly floating on a calm summer lake waiting for a fish to strike their lure.

Giant seadevils are called giants because of their size relative to other seadevils, not because we would regard them as giants relative to us. Northern giant seadevils, *Ceratias holboelli*, are the largest of all seadevils with females reaching lengths of up to 127 cm tip to tail and 86 cm excluding the tail fin (the standard length used by fisheries scientists). They are also the most sexually dimorphic seadevils with females as much as 60 times the standard length and at least 500,000 times the weight of males.[3] This makes *C. holboelli* not only outstanding among seadevils—but the most sexually dimorphic of all known vertebrates. Fortunately for us northern giant seadevils are also one of the best described and widely distributed of all seadevils, occurring in all oceans except the Mediterranean Sea and the southern ocean below about 45° S, where they are replaced by a very similar but smaller species, the southern giant seadevil, *C. tentaculatus*. Although seadevils are difficult to study and poorly known in general, we do know enough about northern giant seadevils to piece together their life history and begin to understand why males and females have come to be so different from each other.

No one has ever studied giant seadevils in their natural habitat, and so everything we know about them has been inferred from detailed studies of dead or moribund specimens brought to the surface, mainly by open water trawls, and by extrapolation from the much better-known shallow-water anglerfishes. All of this knowledge and inference can be found in two wonderfully comprehensive monographs that combine descriptions of each species with scholarly inferences about their life histories, ecology, and mating behavior. The first was written by Danish fisheries biologist and scholar Erik Bertelsen and describes everything that could be discerned from specimens described prior to 1951, the bulk of which were collected during a series of oceanographic cruises by the Royal Danish Research Ship *Dana* in the years 1920 through 1930.[4] The second was written by Theodore W. Pietsch, Professor and Curator of Fishes at the University of Washington in Seattle, and updates the earlier monograph with new information accumulated between 1951 and 2009.[5] From these encyclopedic resources and a little extra input from other scientific articles, I have been able to assemble a reasonably detailed account of the life stories of male and female *C. holboelli*.

Bertelsen's monograph begins with a beautiful illustration of the life stages of male and female *C. holboelli*, and this provides a good starting point for our investigation of sexual differences in this iconic species (figure 8.1). The only catch with this illustration is that you have to pay attention to the scale bars associated with the drawings to appreciate the enormous growth of the females from the larval to adult stage, especially relative to the males. If we were to put the drawing of the adult female at the top of the page on the same scale as that of the larvae at the bottom of the page, we would have to multiply the size of the adult female by fifty or, alternatively, shrink the larvae down to one-fiftieth of their current size. By comparison, we would have to multiply the drawing of the juvenile male by a factor of only 1.2, and that of the juvenile female by a factor of 6 to put them on the same scale as the larvae.

Keeping these relative sizes in mind, consider the adult female at the top of the drawing. She is somewhat less bizarre and more fish-like in general appearance than many other seadevils, but a few peculiarities do stand out. Perhaps most obviously, she is remarkably short on fins. Most ray-finned fishes have two large dorsal fins along their backs, the more anterior of which is supported by stiff, bony fin rays. Our female seadevil entirely lacks the anterior, bony dorsal fin and has only the soft posterior one. Even this is small and sits well back on the body immediately above a similar-sized anal fin on the underside of her body. The paired pelvic fins, which in most fishes sit low down on either side of the abdomen just back of the head, are also absent. A pectoral fin can be seen on the side of the body just behind the head, but it is so tiny that it is dwarfed by the big, deep body. Only the caudal or tail fin is large, but it is weakly webbed and seems not well designed as an organ of propulsion. It is hard to imagine that this odd assortment of fins could generate anything other than rather

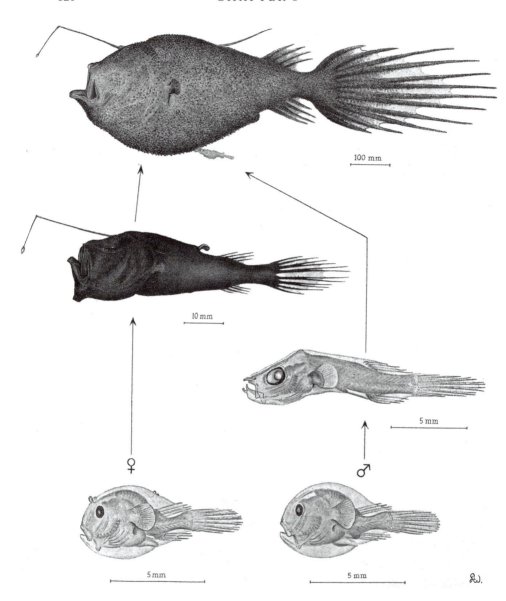

FIGURE 8.1. Life histories of male and female northern giant seadevils (*Ceratias holboelli*). The bottom panels show a larval female (left) and male (right) with 5-mm scale bars. The center panels show a female (left) and male (right) after metamorphosis but prior to sexual maturity (the juvenile stage). Note the 10-mm scale bar for the female and 5-mm scale bar for the male. The top panel shows a fully mature female with an attached, parasitic male. Note the 100-mm scale bar. Drawings by Poul H. Winther, reprinted with permission from Bertelsen (1951).

wobbly, inefficient forward motion. This is hardly a body plan designed for actively pursuing prey. Rather, the morphology tells us that female giant seadevils must spend most of their time drifting, using their fins primarily for orientation and stability rather than for active movement.

A second odd feature of our adult female is that she appears to lack eyes. A hint of an eye is evident in the juvenile female, but nothing hints at an eye in the adult. This is because the eyes of females cease growing when the female larvae metamorphose from the larval to the juvenile stage, remaining no more than 3 mm in diameter even in the largest adults. Further, as the female grows, her tiny eyes gradually sink into the skin and become covered with a transparent layer of skin so they are barely discernable. The structure of the eye itself suggests that it is not capable of forming an image and probably serves only to detect light. All of this would seem to indicate that adult females make little or no use of vision and are essentially blind predators in a black world.

Surprisingly, although almost blind themselves female seadevils have adopted a foraging strategy that depends on the visual capabilities of their prey. Their lure (called the *esca*) hangs at the end of a long supporting spine (called the *illicium*) that extends forward from the foreheads of juvenile and adult females (figure 8.1). The lure is tipped with a bulbous light gland filled with bioluminescent bacteria. This is a highly specialized organ that nurtures the bacterial colony and also regulates the amount of light that can be seen from the outside. Reflective and pigmented layers inside the bulb determine the color and intensity of light, and a tubular light guide focuses the light into a bright beam. Female seadevils can turn their light on and off, creating blinking or wavering patterns that are specific to each species. The highly mobile supporting spine is like an angler's fishing pole and can be pushed back and forth and vibrated, allowing the females to bounce and jiggle their lure, much as a human angler does to simulate living prey and attract the attention of predatory fish in search of a meal. Theodore Pietsch speculates that each seadevil species has adapted the light and movement patterns of its lure to mimic the appearance and motion of the small animals (mostly crustaceans) that would normally be eaten by the seadevil's prey.[5] The lure thus flashes and dances directly in front of the female's mouth, while her dark, hulking shape remains invisible in the gloom. Any animal attracted to this lure is quickly snapped up by the female and impaled headfirst on her rows of dagger-like teeth. Of course this foraging strategy is not without risks for the seadevil herself. Once she bites down on her prize she is apparently unable to let go, at least if the captured individual is large. Dead female seadevils have been found floating at the surface with fishes larger than themselves wedged in their mouths, the two species locked head to head in a suffocating grip that was ultimately fatal to both. The dancing lure may also attract predators that are even larger than the seadevil and capable of gobbling her up along with her lure.

Being a sit-and-wait (or more appropriately "drift-and-wait") ambush predator in the bathypelagic zone is not only risky; it also entails a lot of solitary waiting and not much eating. Animal densities are uniformly low and prey detection distances are short, so encounter rates between predators and prey are low. Female seadevils have adapted to this scarcity in a number of ways. Most obviously, they attempt to attract prey with a glowing lure, rather than relying on chance encounters in the gloom. It is likely that a female can sense the approach of a potential meal by the pressure waves it produces as it moves through the black water.[6] One imagines her waiting motionless in the gloom while she employs her lure to maximum effect until the hapless victim is almost upon her. Then, sensing its proximity, perhaps just as it snaps at her lure, she executes an extremely effective and energetically efficient capture maneuver. This consists of rapidly opening her huge mouth, protruding her toothy upper jaws, and expanding her opercular and pharyngial cavities (essentially her cheeks and throat) so that water rushes in, bringing the hapless prey with it with implosive force. In a flash the prey is impaled on her backward-facing teeth, and her mouth snaps shut. With their huge heads, wide gape, and distensible mouths, throats, and stomachs, adult females are able to handle very large prey. This means that one successful capture can probably provide enough nutrients to sustain an adult female for months. Most of the female seadevils that have been captured in nets or found on beaches have had empty stomachs, which suggests that they endure long fasts between prey captures. During these long intervals they conserve energy by lowering their metabolic rate and floating passively rather than actively swimming in search of prey. By these tactics, Pietsch estimates that mature females probably need only consume prey equivalent to 10 percent of their body weight as infrequently as every three months, and larger meals would permit even longer intervals between catches.

And so life goes for female giant seadevils, drifting, angling, and digesting in perpetual darkness. But what about mating and spawning? The same problems that apply to prey encounter rates apply to encounter rates with potential mates. Seadevils are decidedly solitary. As far as we know they do not form schools where males and females might encounter each other. Nor do they seem to have seasonal migration patterns that might bring the sexes together in high densities. On the contrary, encounters between males and females are probably rare and precious, and failure to find mates may well be a major limitation on the population sizes of these species.[7] For the sexes to have any hope of finding each other, one sex must devote itself to seeking the other, and this role falls to males. While the juvenile females are eating and slowly growing, the juvenile males are fully focused on finding a mate. At metamorphosis, they lose the tooth and jaw morphology necessary for handling prey and so, like our male garden spiders, they are unable to feed while they search. Fueled by energy reserves stored prior to metamorphosis, they swim constantly, peering through the gloom with their huge eyes, searching for the glimmer of a mature female. In many seadevil species the males also employ huge nostrils to scent the water for hints of female pheromones, but male giant seadevils have only rudimentary nostrils and olfactory tissues that are probably not functional. Instead, they have unusually large, bowl-shaped eyes equipped with large lenses and even larger pupils that enhance light-gathering capacity and provide a wide, binocular field of view (figure 8.1). We do not know how male giant seadevils recognize their mates without pheromone cues. However, the color and emission pattern of the light organ (the esca) are species-specific, so it seems likely that they cue into that. Juvenile female giant seadevils also have light organs on their backs. The stout, curved appendage that can be seen on the juvenile female's back just in front of her dorsal fin in figure 8.1 is one of a pair of club-shaped organs called *caruncles*, each of which contains a photophore filled with bioluminescent bacteria. The nubbins of these can be seen even in the tiniest female larvae. The photophores in the caruncles are completely shielded from the outside so that no light escapes, but females seem to be able to release the glowing bacterial fluid into the water, producing a shower of pinpoints of light. Like the eyes, the caruncles do not grow with the female and are small and degenerate in large, mature females. Their prominence in juvenile females suggests that they may serve as shorter-range, ephemeral signals for males, perhaps indicating a female's readiness to accept a mate, but this is pure speculation on my part. I imagine the tiny males wriggling through the black water, endlessly searching for the flickering light from a female's lure and perhaps awaiting a welcoming shower of sparks on close approach.

The juvenile males are clearly adapted for their task of endlessly swimming and searching for a mate. They have large, mobile pectoral fins, smooth skin, long tails, and long, slim, muscular bodies (figure 8.1). They also have relatively small, streamlined heads (less than 30 percent of their standard length) and no externally visible sign of a lure or its supporting spine. They are so small (less than 1.2 cm total length) and so unlike their females, that zoologists initially failed to recognize them as belonging to known anglerfish species. In fact all male seadevils captured during this mate-searching phase were originally classified as a separate family of anglerfishes. It was not until the 1930s that they were recognized as belonging to the same families as female seadevils, and it took another two decades for them to be assigned to the correct genera and species.[8] This is truly sexual dimorphism writ large.

The tiny males do not eat while they are searching for a mate, and so they do not grow at all. They subsist on reserves stored in their livers during their larval stage, and as time goes by their livers gradually shrink. It is likely that the searching phase of their life cycle is short, on the order of weeks or at most a few months. Like our male garden spiders they must find a female before they starve or are eaten (a likely prospect for a small fish in the open ocean). If a male is fortunate enough to find a female, his life takes another dramatic turn. Using specialized, pincer-like denticles that have replaced his larval teeth, he latches onto to the female's belly, usually just in front of her anus. (You can see a small attached male on belly of the mature female in figure 8.1.) Somehow the tiny male manages to avoid being snapped up before grabbing hold. Male-female interactions have never been observed in seadevils, so researchers really do not know what goes on as the male hones in on a potential mate. It may be that he avoids being eaten simply because he is too small for her to consider him as prey, or perhaps he manages to approach stealthily from the rear and latches on before she knows he is there. Certainly avoiding her lure and the huge mouth behind it would be a good tactic. It is even possible that the female recognizes him as a mate and encourages his sexual bite, perhaps in response to species-specific touching by the male. However they manage it, giant seadevil males do succeed in attaching themselves to their mates, and once they have done so, they are there for good.[9] They abandon free living entirely and become parasites on their mates. In the process they transform themselves from lean, streamlined juveniles into plump, spiny adults, hanging from the bellies of their giant mates (figure 8.2). As the male morphs into a true parasite, the tissues of his jaws and snout grow forward and meld with the flesh of the female, while at the same time the female's tissues extend into and around his mouth. Complementary capillary webs form in the connecting tissues of both partners, and the blood vessels become aligned and appear to fuse, so that the two animals seem to share the same circulatory system. Having ceased growing during his free-living stage, the male now resumes growth fueled by nutrients provided by his mate. The parasitic males grow considerably reaching total lengths of 7–19 cm, and the size differential between sexes declines so that the attached males average about a tenth the length of their mates.[10] The eyes of the parasitic males degenerate, and their skin darkens and becomes covered with small spines to match the skin of their mate. Their testes enlarge until, astoundingly, they occupy more than half of the body cavity and form a visible abdominal bulge that changes the body shape from svelte to decidedly plump. Immobile and misshapen, the parasitic males remain attached to their

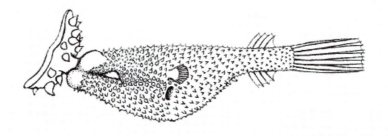

FIGURE 8.2. Parasitic male *Ceratias holboelli* attached to a female. A portion of the female's abdomen can be seen at the left of the drawing. Note the spiny texture of the skin of both male and female, the ring of tissue around the male's mouth, and the papilla extending from the female and projecting into the male's mouth. Reprinted by permission from Regan (1925), figure 3.

mates for the rest of their lives, their only function being to produce sperm and fertilize the female's eggs.

While the attached male is developing his testes and maturing his sperm, the female is reacting to his presence with changes in her own reproductive system. Unlike other fishes neither male nor female seadevils become sexually mature until they acquire a mate. The female's ovaries mature and yolk is added to her eggs only once a male has attached himself. This is another adaptation for survival in an environment where encounter rates with both prey and mates are so low. Prior to mating both sexes remain in a juvenile state without mature gonads so that males are able to devote all their energy to mate searching, while females put their resources into somatic growth. Although males can subsist as free-living juveniles for only a few months at most, females probably remain juveniles for many years, possibly as long as decade.[11] During this time they increase in standard length by about thirtyfold, from only 1.7–2.1 cm at metamorphosis to more than 55 cm by the time they acquire a mate.[12]

Once a juvenile female is prepared to accept a male, she faces a definite shortage of opportunities. Of the 161 metamorphosed *C. holboelli* females that had been cataloged by the time Pietsch prepared his 2009 monograph, only 18 (11 percent) had males attached, so it would seem that the vast majority of females are waiting for males to appear. The fortunate female that acquires a mate can begin to yolk up her eggs in preparation for spawning, but here she faces a second dilemma. She is drifting in the water column, probably 2,000 m (1.2 miles) below the surface and possibly several thousand meters above the ocean floor. When ready to spawn she will have millions of eggs (the one female *C. holboelli* that has been dissected contained 5 million ripening eggs!). Given what is known about anglerfishes that live in shallower waters, it seems likely that female seadevils spawn their eggs in successive batches over several days or weeks.[13] Even so each batch likely contains tens to hundreds of thousands of eggs. Her mate is tiny and anchored in place on her belly. No matter how large his testes, it seems unlikely that he could distribute his sperm widely enough to fertilize all her eggs if they were left to disperse through the water when released. How can she release her eggs and make sure that they will be fertilized? Female seadevils have a solution for this problem. Like their shallow-water cousins they release their ripe eggs in a sheath of mucoid, gelatinous material that initially trails below their bellies and holds the eggs close to the spawning male.[14] The sheath material is riddled with tiny pores that connect each egg to the water. On contact with seawater, the sheath acts like a sponge, sucking water and sperm into each pore and bringing the sperm to the unfertilized eggs. We do not know how long the eggs in their sheath trail after the female once spawning is completed, but eventually the whole mass comes loose and floats up toward the surface. At this stage it is appropriately called an *egg raft* as it carries its cargo of developing embryos up to the light-filled, epipelagic zone where warm temperatures and abundant food await the hatching larvae.

Although adult seadevils live in the bathypelagic zone, their eggs and larvae develop in much warmer and more productive surface waters, most commonly at depths of less than 65 m. The adults seem to be able to time their annual spawning to coincide with favorable conditions for larval development at the surface in spite of the absence of any obvious seasonal cues in their perpetually dark and cold environment. In the north Atlantic, for example, larval seadevils are found in surface waters primarily in July and August, the months when temperatures are warmest and when the planktonic animals on which they feed are most abundant. The drifting eggs hatch into tiny pear-shaped, transparent larvae that already display hints of their future sexual dimorphisms (figure 8.1). Females already have nubbins on their foreheads and backs, presaging the spine that will support their lure and the caruncles from which they will release their showers of sexual sparks. The larvae grow rapidly and begin to metamorphose within a few months, by which time they have reached total lengths of 0.8–1 cm. Female larvae tend to be larger than males by the time they reach this stage, but we do not know if this because they grow faster or because they delay metamorphosis until they have reached a larger size. During metamorphosis both sexes increase in size and take on their adult morphology. Females increase their head and jaw sizes disproportionately, develop their spines and lures, and become darkly pigmented. Males, in contrast, become longer and slimmer, with relatively small heads and with jaws adapted for latching onto females rather than for feeding.

Metamorphosis signals the return journey to the bathypelagic depths for both sexes. By the end of summer of their first year the males have metamorphosed to their free-living, mate-searching stage and have returned to the depths where the older juvenile females are most likely to be found. Because the males do not eat during this phase, it seems likely that they attempt to attach to the first female that they come across. However, the seasonality of egg and larval production indicates that, no matter when the males become attached to their mates, spawning will not occur until the following spring or summer. Thus, males probably live as parasites on their mates for eight or nine months before fertilizing their first batch of eggs. As described above the females follow a somewhat different path. During metamorphosis they descend—but not to the depths most frequented by mature females and males. They seem to spend a juvenile period in slightly shallower water, at depths between 1,500 and 2,000 m (0.9–1.2 miles) where they continue to eat and grow slowly, probably for several years, before descending another 500 m (1,600 feet) or so to the depths where they will mate and spawn.[15]

The iconic giant seadevil females, floating almost motionless in the blackness of the ocean deep with misshapen, parasitic males hanging from their bellies, represent the most extreme sexual dimorphism in size, form, and lifestyle among all vertebrate animals. The two sexes are readily distinguishable in the earliest larval stages, differ noticeably in size by the time metamorphosis begins, and by the time metamorphosis is complete, are so different that for many years they were not even recognized as belonging to the same taxonomic families, let alone species. The tiny, free-living males are sleek swimmers with huge eyes and look nothing like the spine-covered, hulking females with their massive jaws and seemingly eyeless faces. The size differential between the free-living males and their potential mates is staggering, females apparently being up to 60 times longer and weighing up to 500,000 times more than males. Even once they become parasites the males average only a tenth as long as their mates and probably less than one-thousandth the weight. One wonders why seadevils have adopted such a bizarre mating arrangement. Unfortunately, we cannot answer this question by observing live seadevils in nature or in aquaria, so there is no way to directly measure the effect of these behaviors and morphologies on the Darwinian fitness of each sex. However, we can surmise what is likely going on by looking at females first and considering in what ways the females are likely to benefit from being so large. Being fishes, their fecundity no doubt increases steeply with body size, and so larger females likely produce more eggs per season and per batch.[16] Being gape predators that engulf and swallow their prey whole, larger females also benefit from being able to process larger prey. In the deep sea this is particularly advantageous because encounters with potential prey are so infrequent that females must be able to survive for long periods between meals. Large size probably also enables females to eke out their stored energy for longer periods between meals because mass-specific metabolic rate declines as body mass increases. Being large also reduces the chances that the females will themselves be eaten by larger predators, a risk that is increased by their tactic of attracting their own prey with a highly visible lure. A final advantage of larger size derives from the parasitic mating system itself. Larger females can presumably support the energetic and nutrient requirements of their attached males with less sacrifice to their own growth, maintenance, and fecundity. They may even be able to allocate more nutrients to their parasitic males thereby allowing them to grow larger and produce more sperm. Some support for this latter hypothesis comes from comparing the sizes of females and their attached males: the smallest attached male (standard length 3.5 cm) was found attached to the smallest female, and there is an overall positive correlation between the sizes of the males and their mates.[17] Unfortunately, without studies of live animals in the wild, we cannot determine the relative importance of these various size advantages for female giant seadevils, nor can we be certain that other factors do not also come into play. Nevertheless, it is clear that the benefits of increased size have combined in some fashion to produce giant seadevil females that, in addition to being much larger than their males, are also the largest females of any seadevil species.

Male seadevils are no doubt small for many of the same reasons that male yellow garden spiders and blanket octopuses are small. Because mature females are so rare and difficult to find, these males focus first and foremost on finding mates. As we saw in earlier chapters larger males tend to have the advantage when competing for mates by either direct physical contests or competitive displays to attract females, and hence males tend to be large in species where these types of competition occur. However, in the absence of these forms of male sexual competition, there is little reason for males to be large. On the contrary in species where females are widely dispersed and male fitness depends strongly on simply finding a mate, smaller males seem to have the advantage. Like yellow garden spiders and blanket octopuses, male northern giant seadevils have evolved a morphology and life history that maximize their probability of encountering a mate and of being ready to mate with her whenever such an encounter occurs. By metamorphosing to their adult form at an early age, the seadevil males minimize their chances of dying before they can begin their search for a mate, increase the proportion of their lifespan devoted to that search, and ensure that they will be ready to mate when and if they do encounter a female. Like male garden spiders seadevil males forgo eating entirely during their mate search phase, and running out of fuel before finding a mate is a clear risk. This means that during their larval stages they have to divert energy to stored fuel rather than to overall somatic growth, and so they metamorphose at a smaller somatic size than females. The smooth, streamlined shape and powerful tail of metamorphosed males enable them to swim strongly and much more efficiently than females, and these characteristics, along with their large eyes, are clearly adaptations for mate search. Whether their small size per se is adaptive at this stage is less clear. As a general rule swimming speed increases with size in fishes. In addition if energy reserves scale at least proportionally with body size, larger fish should be able to go longer without food because of their lower metabolic rate per unit mass. These relationships suggest that male seadevils might be better at finding mates if they were larger. If this is so, the reason they remain small is no doubt because being larger would have negative impacts on other components of their fitness. For example they would have to spend more time growing before searching for mates, and hence more would die before even beginning their search. Problems would also arise for larger males once they encountered a female because she might not be large enough to support them once they transitioned to their mature, parasitic phase. In addition females that have to divert more energy to maintaining a large male will necessarily have less energy for their own growth and for producing eggs. Larger males might thus have lower lifetime fertility even if they did find a willing mate. A small mate with large testes fits the bill much better. Small males can attach to even the smallest females available and then grow once attached in accordance with the energy that the female can provide. It is a system that tailors the size dimorphism of mated pairs to the foraging success and growth of the females. Even if larger males were more efficient at finding females, the parasitic mating system clearly favors males that are very much smaller than their mates. Not surprisingly, both the smallest males (in the family Linophrynidae) and the greatest sexual size dimorphism (in the family Ceratiidae) occur in seadevil species where males are obligate parasites on their females.

The parasitic mating system has an additional implication for sexual dimorphism in seadevils. Once males and females have paired up, they remain attached for the remainder of their lives. Of the eighteen female *C. holboelli* that have been found with males attached, only one had two males. Thus, the norm for this species seems to be one male per female. The lone male will fertilize repeated spawnings by his mate during a given spawning season, and the pair will continue to spawn in successive years until death. Although we do not have lifespan data for *C. holboelli*, commercially exploited anglerfishes of similar sizes are known to live as long as sixteen to twenty-four years.[18] It thus seems likely that the attached male and female *C. holboelli* will spawn together for ten years or more, barring early death by misadventure. Thus, unlike male garden spiders and blanket octopuses, which die after a single mating (in spiders after having achieved two insertions), giant seadevil males lucky enough to find a mate are likely to spend by far the greatest portion of their lives as parasitic males. Their reproductive success is not based on one fatal episode, nor are they likely to be competing with the sperm of any other males. As a consequence they have become highly specialized for their lives as parasites, and, as I have suggested above, adaptation to the parasitic lifestyle may well explain why they are so much smaller relative to their mates than even our spiders or octopuses. As we see in the following chapters males that rely on repeated spawnings but forgo eating altogether once they have become sexually mature are smaller still.

CHAPTER 9

Bone-Eating Worms

Female Tubeworms with Harems of Minuscule Males

So far I have described examples of extreme sexual differences in animals that live fully on land (great bustards and garden spiders), partly on land and partly in the sea (elephant seals), in freshwater (shell-carrying cichlids), in the shallow waters of the open ocean (blanket octopuses), and in the vast expanses of the ocean depths (seadevils). In this chapter we descend to the bottom of the ocean to examine the lives of animals that make their living on the sea floor itself. The species described in this chapter, bone-eating worms, come from the segmented worm phylum (Annelida), which also contains more familiar wormy types such as earthworms and leeches. However bone-eating worms differ from these distant cousins in many ways, including major differences in their adult body plans, patterns of growth and development, and ways of making a living. The most obvious difference is that they are sedentary animals that remain anchored in one spot throughout their adult lives, a lifestyle biologists call *sessile*. Being anchored to the bottom and hence immobile presents a particular difficulty for animals with separate sexes. How can the sperm get to the eggs if the adults of one sex cannot seek out the other for mating? By far the most common solution to this problem is to somehow disperse the sperm so that they are able to encounter the eggs at some distance from the male parent. On land a familiar example of this strategy is pollen dispersal by flowering plants and conifers. The tiny pollen grains reach the female ovules by relying on external forces such as wind, water, or transport by insects, bats, or birds. Most sessile animals living in aquatic habitats accomplish the same goal by releasing their sperm into the water and allowing chance and currents to carry the sperm to waiting eggs. The eggs are usually retained by the mother and released only after they are fertilized, but some species release both sperm and unfertilized eggs so that fertilization occurs in the water column.[1]

Bone-eating worms are among the few dioecious, sessile animals that have devised a way of getting their eggs fertilized without requiring their sperm to make an obviously chancy open-water swim, and they have accomplished this by evolving one of the most extreme sexual dimorphisms in the animal kingdom. The animals that we see in colonies of bone-eating worms are all sessile adult females. To find the males we have to look inside the transparent tube that surrounds each female, forming a stiff, protective outer wall. Harems of dwarf males can be found clinging to the inner surface of these tubes. They fasten themselves onto the female's body as larvae, rapidly mature into minuscule functioning males, and proceed to spend the remainder of their short lives making sperm to fertilize her eggs. The tiny suitors live on their female's body, but, unlike seadevil males, they subsist entirely on their own stored reserves. Both sexes benefit by this arrangement. The males are protected and are able to release their sperm close to the female's unfertilized eggs, and the females have a ready supply of sperm. The tiny, nonfeeding males are neither parasites nor freeloaders. They are simply partners providing sperm at no apparent cost to their mates. The extreme reductions in size, morphology, age at maturity, and lifespan of these males are really just extreme versions of the adaptations we saw in orb-web spiders, blanket octopuses, and seadevils. Nevertheless, bone-eating worms add new twists to this story, and their tale is well worth telling.

Bone-eating worms are relatively new additions to the encyclopedia of animal life. In February of 2002 a group of marine biologists working with the remotely operated submersible *Tiburon* from the Monterey Bay Aquarium Research Institute (MBARI) discovered a decaying gray whale carcass on the seabed at a depth of 2,893 m (1.8 miles).[2] "Whale 2893," as the carcass came to be known, was resting on the bottom of Monterey Canyon, about 31 km (19 miles) off the central coast of California. It was a relatively recent casualty with flesh and soft tissue still clinging to the skeleton, and it was teeming with life. Among the many animals swarming on and over the mass of flesh and bone were colonies of tiny, sessile worms, densely packed on patches of exposed whale bone. These unusual creatures were deeply anchored in the whale bones. Their thin, vermiform bodies reached upward through delicate, transparent tubes and were topped with elegant crowns of four bright red, feathery tentacles or *palps*. Among the biologists studying the material from that voyage were two scientists from MBARI, Robert Vrijenhoek and Shana Goffredi and a third, Greg Rouse, then from the South Australian Museum and later at Scripps Institute of Oceanography in La Jolla, California.[3] Recognizing that these strange worms were something that had not been seen before, the three researchers set themselves the task of determining what kind of animals they were and where they fit into the grand phylogeny of the animal kingdom. Two years later they published the first description of an entirely new animal genus in the prestigious journal, *Science*.[4] They called their new genus *Osedax,* meaning bone-devourer, and gave the included species the common name of bone-eating worms.

Genetic analyses placed the new genus within the phylum Annelida, Class Polychaeta, and family Siboglinidae, the marine tubeworms.[5] This family contains large tubeworms found at deep-sea hydrothermal vents and cold seeps (called vestimentiferan tubeworms), long, thin tubeworms characteristic of muddy, anoxic deep-ocean bottoms (called frenulate tubeworms), and a few species in the genus *Sclerolinum* that live on decaying organic matter such as wood or rope. All share the characteristic of living in close-fitting tubes made from opaque chitin or transparent mucous-like material and firmly attached to a substrate. Although the new *Osedax* worms had some similarities to the other tubeworms, the research team quickly identified unique aspects of their adult morphology, biology, and ecological niche. Most relevant for us is that the colonies of bone-eating worms on whale 2893 seemed to be composed entirely of females. Male and female tubeworms are typically very similar in appearance and occur together, often in dense colonies. In some species experts can distinguish males from females by the position of their genital pores, and females may be slightly larger and thicker-bodied than males, but otherwise there is little to distinguish the sexes other than their internal gonads. However, the males were missing from the colonies of *Osedax*. Only when the researchers removed the females from the bones and examined them under a microscope did they discover the missing males. Masses of tiny, larva-like males were found clinging to the inner walls of the females' tubes. This extreme sexual dimorphism was so surprising that the team entitled their ground-breaking *Science* paper: "*Osedax:* Bone-eating marine worms with dwarf males," giving equal billing to the unique capacity of their worms to subsist on bone and to the presence of dwarf males.

Rouse and colleagues distinguished two species of *Osedax* in the first samples from whale 2893.[6] They named the larger species *rubiplumus* (*rubi* for red and *plumus* for feather), a name that alludes to the four feathery red palps at the anterior end of each worm. The second, somewhat smaller, species was named *frankpressi,* in honor of Dr. Frank Press, a distinguished American geophysicist who had among many other accomplishments served for twelve years as president of the National Academy of Sciences. These two species made their debut together in the 2004 *Science* article. In subsequent years thirteen more *Osedax* species were discovered on whale bones in Monterey Canyon, at depths ranging from 382–2,893 m (1,253–9,821 feet), with as many as seven species on a single carcass.[7] Two additional species were discovered at depths of only 30–250 m (98–820 feet) off the coasts of Japan and Sweden, so that by 2010, the genus *Osedax* contained at least seventeen species.[8] Although *Osedax* worms will readily colonize the bones of other vertebrates ranging from cows to large fishes when these are placed on the seafloor, they are primarily whale-bone specialists, and their dense colonies are one of the major agents of decomposition of whale carcasses.[9]

When observed alive the most visible and striking feature of almost all *Osedax* colonies is the delicate crown of palps sported by each female (plate 17).[10] Long and graceful, the palps are actually free-floating gills that serve to exchange oxygen and carbon dioxide with the surrounding seawater. The crown extends from the top of a long, thin, whitish trunk enclosed in a transparent tube. Both the trunk and the palps can contract so that the animal is able to withdraw at least partially into its tube, and only the tube, trunk, and crown are visible in live worms. The rest of the worm's body is embedded in the whale bone, and it is in these hidden regions that the truly bizarre biology of bone-eating worms becomes apparent. The trunk terminates in a large, bulbous ovary, the egg-production factory, which becomes swollen with eggs as the female matures, and an irregular system of roots extends from the base of the trunk through the bone. The ovary and roots are sheathed in a greenish layer of tissue that lies just beneath the outer layer of skin, and this green tissue, called the *trophosome,* is the key to *Osedax* biology.[11] Like all siboglinid worms, *Osedax* do not have a mouth or a gut and so are unable to ingest food directly. In spite of being called bone-eating worms they do not "eat" bone in the conventional sense. Instead, they rely on a unique, mutually beneficial relationship with colonies of large, rod-shaped bacteria that are able to digest collagen from the bones.[12] Juvenile female *Osedax* somehow acquire these bacteria from the surrounding seawater and sequester them in specialized cells called bacteriocytes embedded in the trophosome tissue. Blood vessels running down the female's trunk from the palps to the trophosome provide the bacterial cultures with oxygen and remove the waste carbon dioxide. Thus nurtured and sheltered, the bacteria grow and multiply inside the trophosome. The host female then obtains her nutrition secondarily from the bacteria, most probably by digesting the bacteriocyte cells when they become replete with bacteria. The obligate dependence of *Osedax* females on their bacterial cultures and vice versa is a form of biotic interaction that biologists call a *symbiotic mutualism.* Other siboglinid worms also obtain their nutrition through symbiotic mutualisms with bacteria, but in those relationships the bacteria obtain their energy by oxidizing methane or hydrogen sulphide dissolved in the seawater. The partnership between bone-eating worms and bone-digesting bacteria is unique in the animal kingdom.[13]

Most of what we know about bone-eating worms comes from the colonies found on whale carcasses in Monterey Canyon. After the initial discovery of *Osedax* on whale 2893, the scientists at MBARI deliberately sank five more whale carcasses in the canyon, at depths ranging from 382 m to 1,820 m (1,253–5,972 feet), for the express purpose of studying the organisms that colonized the carcasses. They commandeered carcasses that had beached along the nearby coast, towed them out to sea, weighed them down with iron train wheels so that they would sink, and released them at prescribed locations. Over subsequent months and years they used the MBARI fleet of remotely operated submersibles (ROVs) to follow the colonization and degradation of each carcass. The research team studying *Osedax* grew to include a diverse group of scientists from other laboratories in California, Oregon, and Denmark, and by 2010 they had described a diverse community of *Osedax* in Monterey Canyon.[14] The various species differ somewhat in size, shape, and color (table 9.1) and in the depths and ages of the whale carcasses that they colonize, but they share the same adaptations for extracting nutrients from the bones of dead whales.[15]

All *Osedax* start life as fertilized eggs released into the water by their mothers. The females extrude their eggs one by one from the end of a long, tubular oviduct that extends from the ovary up through the trunk and terminates amid the crown of palps. When a female spawns, her batch of eggs can be seen moving in line up the oviduct and popping out the top, much like peas out the end of a peashooter (see the middle panel of plate 17). The eggs are heavier than water and most probably sink down among tubes of the colony. In this relatively sheltered environment the eggs start developing within hours, and within two days they have developed into tiny ovoid larvae.[16] At this stage they look like typical annelid larvae, which are called *trochophore* larvae because they propel themselves using a ring of beating, hair-like *cilia* called a *prototroch*. The larvae do not eat but instead rely on yolk provided by their mothers, and to this end each larva has a large cell packed with yolk granules. Beating their tiny cilia they twirl and drift at the mercy of the currents, sustained by their supply of yolk. This is the dispersal stage for *Osedax* and their only chance to find a suitable carcass on which to settle. However, despite the critical importance of this goal, they seem to retain their dispersal capability for a

TABLE 9.1.

Main distinguishing characteristics of three *Osedax* species

		O. rubiplumus	*O. frankpressi*	*O. roseus*
Females	Maximum trunk–crown length[a]	59	23	26
	Trunk length[b]	38	4.5	7
	Palp color	Brilliant red	Red, each with two longitudinal white stripes	Bright red, each with two longitudinal white stripes
	Roots	Long, branched	Robust, lobate	Long, branched
	Tube	Firm, transparent	Gelatinous, transparent	Gelatinous, transparent
Males	Maximum length	1.1	0.25	0.21
	Average number per female	17–26	—	3.5
	Maximum number per female	607	80	14
Eggs	Length × width	0.15 × 0.12	0.15 × 0.12	0.13 × 0.09

SOURCES: Rouse et al. (2004, 2008, 2009), Braby et al. (2007), Vrijenhoek et al. (2008), Lundsten et al. (2010).

NOTE: Lengths and widths are in millimeters.

[a] Measured when contracted in preserved specimens. Palps are much longer when living.

[b] Measurements from single type specimens.

remarkably short time, settling to the bottom within about two weeks.[17] This short dispersal phase is surprising, given that whale-falls and other large vertebrate carcasses must occupy a vanishingly small proportion of the ocean floor,[18] and bone-eating worms require such carcasses to grow and reproduce. Surely the vast majority of *Osedax* larvae must perish without finding a suitable resting place. Such a strategy seems improbable, but it succeeds in the same way that a lottery does: even though the chance of winning may be less than one in 10 million for any given ticket, there are always winning tickets. In an analogous way the rapid colonization of new whale-falls by *Osedax* tells us that the supply of drifting larvae must be great enough to overcome the seemingly insurmountable odds of any given larva finding a suitable landing site.

Once the larvae land on a suitable whale bone, they are able to cling to it using eight pairs of tiny hooks on their posterior end. They then rapidly metamorphose into juveniles and grow quickly into mature females. The first whale carcass sunk by the MBARI crew (whale 1018, named for the depth where it lay) provides an informative time series of this initial colonization because it was visited five times within the first nine months after it was sunk.[19] Patches of *Osedax roseus* were already visible when the carcass was first revisited, less than two months into the experiment. The first bone samples were retrieved three months after the carcass had been sunk, and by then 28 percent of the *O. roseus* females contained eggs in their oviduct, and 27 percent had at least one dwarf male clinging to the inside of their tube (figure 9.1). The density of females had already reached its maximum of about 1.6 females per cm,[2] and the average size of the females (measured as crown-trunk length of preserved animals) had stabilized at slightly more than 1.1 cm. The density and average size then changed little until the sample taken nine months after deployment when the density declined slightly, and increased proportions of smaller females without eggs or mates suggested recruitment of a second generation. This time series indicates that *Osedax* can very rapidly colonize newly available carcasses. Exposed patches of bone quickly fill with firmly-anchored females that grow rapidly to their adult size and begin producing eggs and accumulating mates. Once established the colony is maintained by

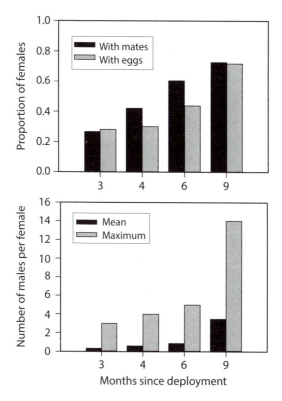

FIGURE 9.1. Change in sexual maturity and sex ratio of *Osedax roseus* over time following the sinking (deployment) of the carcass of whale 1018. Top panel: proportion of females with at least one male (black bars) and proportion with eggs in their oviduct (gray bars). Bottom panel: mean (black bars) and maximum (gray bars) number of males per female. Data are from Rouse et al. (2008).

continuous recruitment of new larvae. Repeated surveys of the whale carcasses in Monterey Canyon have revealed that *Osedax* populations can persist for years: populations of *O. roseus* were still thriving on whale 1018 more than four years after the carcass was deployed, colonies of *O. rubiplumus* were still abundant three years after deployment of whale 1820, and the colonies of *O. frankpressi* on whale 2898 persisted for almost six years and disappeared only when whale bone fragments were no longer visible at the site.[20] However, the longevity of each population is ultimately determined by the degradation of the carcass on which it lives, and every population must eventually go extinct as its carcass disappears. The worms literally eat themselves out of house and home.

To succeed on such ephemeral resources *Osedax* females have to grow rapidly, mature quickly, and spawn prodigious numbers of eggs provisioned with large quantities of yolk.[21] Because the odds of any one of her offspring finding a suitable settling site are so low, the Darwinian fitness of a female depends on production of tens of thousands of eggs. A typical response to such selection would be production of large numbers of very small eggs, but female bone-eating worms cannot rely on this strategy. They must also provision each egg with sufficient yolk both to sustain the nonfeeding larvae until they settle and to support those that turn into males through their entire lives. Producing vast numbers of large, yolk-rich eggs is a tall order for the females and requires enormous investment in reproduction. To that end, the large, bulbous ovaries of *Osedax* occupy an astounding portion of their bodies (plate 17). Within two or three months of settling on exposed bone, the females begin to release their eggs, and from that time they spawn continuously, releasing an average of several hundred eggs per day.

Osedax males also mature quickly and devote a very large portion of their energy and body mass to gamete production, but they take this strategy much further than the females do.[22] As larvae they position themselves along the inner wall of a female's tube, anchor themselves in place with their posterior hooks, and with no further ado initiate spermiogenesis (figure 9.2). Although clearly reproductively mature, they retain their small size and larval morphology, in most species attaining maximum lengths of less than a quarter of a millimeter (table 9.1). The minuscule males contain little else than spermatids (cells that become sperm), mature sperm, and yolk (figure 9.2). Within any given harem the smallest (and presumably youngest) males are typically roundish, full of yolk, and without sperm; larger males are elongate, with little or no yolk, and lots of sperm; and the largest are devoid of both yolk and sperm, having exhausted their resources and reproductive capacity. These males are actually taking the yolk provided by their mothers in the egg and turning

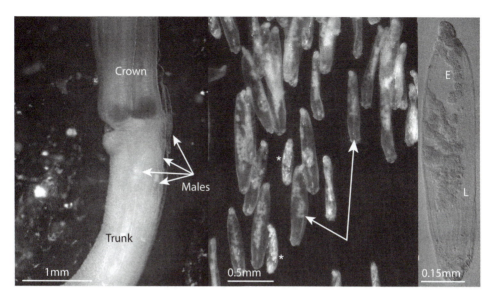

FIGURE 9.2. *Osedax rubiplumus* males. Left panel: The upper part of the trunk of a female and bases of her crown showing part of her harem of dwarf males. The males are attached to the inner wall of her transparent tube, which is barely visible in the photograph. Middle panel: Males of assorted sizes sampled from the female's harem. The smallest males (*) are short and full of yolk. Large males (arrows) contain sperm and spermatids but are largely devoid of yolk. The other males are intermediate in size and contain both yolk and spermatids. Right panel. A single male with early (E) and late (L) spermatids and no apparent yolk. Photographs courtesy of Greg Rouse (Scripps Institution of Oceanography).

it into sperm to fertilize the eggs spawned by their mate, a truly extra-ordinary curtailment of the male life cycle.

Because each male is so small and short-lived, a single mate can clearly not fertilize all the eggs a female is capable of spawning in her lifetime.[23] To achieve her maximum fertility a female has to acquire new males successively as she ages, and she probably also needs more than one male at any one time. Acquiring a sufficient number and continuous supply of mates is a problem for females, and there is good evidence that lack of males restricts female reproductive success. One indication of this is the presence of females with no males in their tubes but with eggs in their oviducts ready to be spawned. This was true for 4 percent of the *O. rubiplumus* females collected from whales 1820 and 2893 and for 10 percent of *O. roseus* females collected from whale 1018. Another indication is the relatively low fertility of the eggs actually spawned by living females. Direct measurement of egg fertility requires collecting the eggs as they emerge from the oviduct and observing them to see if they begin to develop into larvae. This cannot be done in wild populations, but Greg Rouse and his colleagues were able to collect eggs from wild-caught *Osedax* females spawning in laboratory tanks.[24] Only 63 percent of these females had full or nearly full fertility, and 25 percent spawned only unfertilized eggs. These results certainly suggest that males and their sperm are a limiting resource for females. Males are particularly rare early in the life of the colony (figure 9.1), and it seems likely that female fertility is often constrained by lack of males early in their life cycle. However, as time passes, more and more males accumulate until eventually the sex ratio becomes highly biased toward males, almost all mature females have at least one male, and some females have tens or even hundreds of males in their tubes (figure 9.1, table 9.1).[25]

The process of mate acquisition seems to be largely random, which means that some females accumulate more males than others just by chance. By chance some become reproductively mature before they happen to get a mate, as noted above, and conversely, about one in six acquire males before they have ripe eggs for spawning.[26] As time passes and the females get older the probability of being without a male declines, and the mean number of males per female increases.[27] These temporal trends place a premium on female longevity over and above the normal benefits associated with increasing age and size in female animals. As in most animals the lifetime fecundity of female *Osedax* will increase as they age simply because the total number of eggs they have spawned increases with every spawning event. *Osadex* also grow, and so older females are on average larger. If fecundity increases with size, as it does in most animals, older females are likely to spawn more eggs per day in addition to having higher cumulative fecundity. These are the familiar advantages of long reproductive lives and large size that apply to females in most animal species and are no surprise. However, for *Osedax* females long life has the additional advantage of bringing increased egg fertility. When they first mature, most females probably suffer reduced fertility because they have no mates or too few mates to fully fertilize their eggs. However, if a female lives long enough she is likely to accrue a large enough harem to ensure that all of her eggs are fertilized. Female *Osedax* thus pay a fertility cost for early spawning and reap a fertility benefit for spawning later in life.

Given that male *Osedax* only settle in the tubes of established females, it makes sense that the initial colonists of whale bones are always female and that the rise of male densities lags behind that of females. However, if potential colonists are constantly raining down on the whale carcass from the vast pool of drifting larvae, as seems to be the case, what happens to all the males that must land before there are females available? Do they simply perish? Mercifully, the way in which *Osedax* achieve their sex-biased pattern of colonization seems to be much less wasteful than the early death of hordes of male larvae that land prematurely on barren scraps of bone. In fact there appear to be no hordes of male larvae at all. The available evidence suggests that most *Osedax* larvae are neither male nor female but instead have the potential to develop into either sex, depending on the substrate they encounter on settling. Those that settle on bone become females, whereas those that settle on females become males. This form of environmental sex-determination is akin to what occurs in the green spoon worms briefly described in chapter 2.[28] Like bone-eating worms, green spoon worms have extreme sexual dimorphism, and the females accumulate harems of dwarf males. The larvae disperse by drifting, just as bone-eating worm larvae do, and they metamorphose into males or females only after they settle. Almost all larval green spoon worms that land on the ocean floor metamorphose into females that grow slowly and take about two years to mature into large, sac-like adults. In contrast, larvae that land on females quickly metamorphose into males, grow little, and mature within a few weeks. The male developmental pathway is triggered when larvae encounter a chemical or chemicals secreted by mature females, and only about 2 percent of larvae become males in the absence of this trigger. Free-swimming larvae are strongly attracted to mature females and preferentially settle on the female's proboscis. This preferential settlement combined with rapid masculinization of the developing larvae results in the accumulation of harems of males on females and creates strongly male-biased sex ratios in the adult population. The experiments necessary to verify that a similar mechanism of environmental sex determination is operating in *Osedax* have not been conducted, and so we cannot say for certain that this is what is happening during the colonization of whale bones. However it seems the most likely explanation for both the sex-biased pattern of colonization and the absence of any sexual differentiation in *Osedax* larvae prior to settlement.[29]

This brings us full circle in the story of bone-eating worms, from the fertilized eggs popping out the tip of the female's ovipositor, through colonization of newly available bones by undifferentiated larvae, to spawning of mature females with harems of dwarf males clinging to the walls of their tubes. These are bizarre animals by any standards. As adults the females are unlike any other known animals. Only their larval characteristics tie them to other annelid worms, and their symbiotic relationship with bone-digesting bacteria is unique in the animal kingdom. The males are reduced to little more than sperm-producing larvae still fueled by yolk from their mother, and perhaps most amazingly the differences between the adult males and females are probably produced without any underlying genetic differences. The developmental switch that causes a larva to become a male or a female is most likely triggered by an environmental cue: land on bone and become female; land on female and become male. The masculinized larvae simply stop development and start making sperm, remaining larva-like in appearance their whole lives, while the feminized larvae anchor on the bone, acquire colonies of bone-devouring symbiotic bacteria, and proceed to grow trunks, roots, crowns, and ovaries that swell with enormous numbers of eggs. It is hard to imagine any other example that could top this one, and yet I do have one more example to describe. The shell-burrowing barnacles that we meet in the next chapter are not as dimorphic in size as bone-eating worms, nor is their lifestyle as unusual. Nevertheless, with respect to the morphology and behavior of their dwarf males, they are equally extraordinary.

CHAPTER 10

Shell-Burrowing Barnacles

Sac-Like Females with Harems of Phallic Males

Barnacles are arthropods, members of that vast phylum comprised mainly of active, many-legged, hard-bodied animals including insects, spiders, crabs, shrimps, centipedes, and millipedes. Unlike most other types of arthropods, however, barnacles are almost all sessile, bottom-dwelling animals that remain anchored in one spot throughout their adult lives. Acorn and gooseneck barnacles are the most familiar barnacle forms and are often found in very high densities on rocks, pilings, ships' hulls, and even the skins of whales. At low tide you can often find solid mats of these barnacles completely covering the underlying rocks or pier pilings to which they have cemented themselves. Because of this habit of forming a second surface over the underlying substrate, ecologists call acorn and gooseneck barnacles "encrusters." The main difference between the two forms is that acorn barnacles cement their shells directly to the substrate on which they settle, whereas gooseneck barnacles support themselves on fleshy stalks cemented to the substrate and have their shelly plates clustered at the tips. At first glance most barnacles seem more similar to a sessile, shelled mollusk than to their more active arthropod cousins, but on closer examination their arthropod traits become evident. The easiest way to envision a barnacle is to imagine a tiny shrimp-like animal lying on its back and scooping food into its mouth with its legs. Surround that animal with a series of shelly plates and stick it to a rock, and you have a typical barnacle. The name of the arthropod clade to which barnacles belong, Cirripedia,[1] means curl-footed in Latin and refers to the long, feathery legs that protrude from the top of the shells when the animal is feeding. The graceful legs gently curl in and out, harvesting drifting plankton and delivering it to animal's mouth, while the rest of the animal remains safely hidden inside its shells.

Although most barnacles are hermaphroditic, they generally refrain from self-fertilization and instead seek matings with their surrounding neighbors. This is could be problematic given that each member of the amorous pair is anchored firmly in place, but barnacles have worked out a quite spectacular solution. They are distinguished by having the longest penises, relative to their body size, of any known animals. Some have penises that can inflate to lengths exceeding eight times the diameter of their bodies![2] These giant mating organs are kept modestly deflated and furled until the urge to execute a male mating attempt arises. The penis then does all of the hard work of searching for a mate as well as intromission and ejaculation. It stretches itself up and out of the shell and then probes slowly all around the shell, clearly sensing the presence and reproductive state of any individuals within reach. Any neighbors that have mature eggs ready for fertilization will be waiting with their shells slightly open and their kicking feet briefly at rest to allow the penis to enter and deposit sperm. Neighbors not so disposed will shut their shell apertures abruptly on the seeking penis, making their message abundantly clear. At the peak of the reproductive cycle, colonies of barnacles are alive with seeking penises, dancing, probing, and stretching what seems like impossible distances from their anchored bodies.[3]

As daring as this mating tactic is, it has clear limitations when barnacle densities are low. If there are no neighbors within penis striking distance, a barnacle cannot mate as either a male or a female unless it fertilizes itself, which barnacles seem very reluctant to do. The somewhat surprising solution to this dilemma has been for barnacles that tend to occur in low densities to abandon pure hermaphroditism in favor of having males that seek out and settle on larger, sessile females or hermaphrodites.[4] The males are typically very small and have greatly simplified bodies. Once they have found a mate they stick to her for life and devote themselves entirely to fertilizing her eggs. Charles Darwin was the first to describe this curious mating system in barnacles, and his original terminology has stuck. He called the tiny males *complemental* if they paired with hermaphrodites and reserved the term dwarf for males in dioecious species. In *The Origin of Species* Darwin marveled at the morphological degeneration of these males, describing them as "a mere sack, which lives for a short time and is destitute of mouth, stomach, and every other organ of importance, excepting those for reproduction."[5]

Darwin's characterization of dwarf and complemental male barnacles as degenerate is biologically correct. Unlike male *Osedax*, which simply cease developing and start reproducing while still retaining their larval morphology, male barnacles go through a series of larval stages with highly complex and well-developed organs and organ systems before metamorphosing into their simplified, sac-like adult forms. This simplification of adult male form and function illustrates one of the great benefits of true metamorphosis (literally, a remodeling of the body between life stages). Metamorphosis allows each life stage of the animal to be adapted to its own environment without being constrained by the requirements of earlier or later stages. (Think of caterpillars metamorphosing into butterflies or maggots into flies.) Most barnacles have mobile, free-swimming larvae that metamorphose into sessile adults, permanently affixed to a substrate. The requirements of the larval and adult life stages are clearly different. Larval barnacles require complex morphology and sensory capabilities to be able to disperse and eventually find and settle on a suitable substrate, and their complexity is an adaptation for these functions. Metamorphosis allows the typical sessile adults to be adapted for their lives as suspension feeders, anchored in place on a solid substrate. The same metamorphosis frees the complemental and dwarf males to remodel themselves into stripped-down specialists for fertilizing eggs. The tiny males are degenerate forms, as Darwin surmised, but they are nevertheless highly specialized for their function in life.

The typical barnacles I have described so far comprise the by far largest taxonomic group of barnacles, collectively known as the Thoracica.[6]

Although burrowing barnacles share many characteristics with thoracicans, they form a separate clade of their own, called the Acrothoracica. The most obvious distinguishing feature of acrothoracican barnacles is that they lack shells of their own and instead burrow into the calcareous shells and skeletons left behind by other marine animals. The only evidence that the abandoned shells host burrowing barnacles are tiny apertures, generally less than 1 mm long, marking the burrow openings. The barnacles themselves are tiny, most being less than 5 mm in longest dimension, and they live their adult lives permanently affixed inside their burrows. Rather than living in dense colonies with neighbor touching neighbor, they live isolated lives, separated from their neighbors by the calcareous matrix of the host shell. Even in shells with relatively dense concentrations of a given species, the distances between burrows preclude neighborly penile intrusions, and so the typical barnacle strategy of mating with neighbors is doomed to failure. To overcome this limitation burrowing barnacles have abandoned hermaphroditism in favor of a dioecious mating system in which small, structurally and functionally reduced dwarf males cohabit with much larger females within the female's burrow. The males are truly minuscule, averaging less than 0.5 mm long, and most consist of little more than an enlarged testis and a penis that, as one would expect of a barnacle, can be extended to many times the body length (figure 10.1). The males attach themselves permanently to the outside of the female's body, just as other barnacles attach to inanimate substrates, and they then employ their super-sized penises to navigate the distance between their settlement site and the openings of the female's oviducts where fertilization occurs.[7]

The burrowing barnacle depicted in figure 10.1 goes by the scientific name of *Trypetesa lampas*. This species was the first acrothoracican barnacle described (by Hancock in 1849) and remains one of the best known.[8] The adults occur at relatively shallow depths (<50 m) along the coasts of western Europe, the Mediterranean Sea, and eastern North America, and they are most often found burrowed in the calcareous shells of dead whelks and conches that have been secondarily occupied by hermit crabs.[9] As neither sex remotely resembles any familiar animal

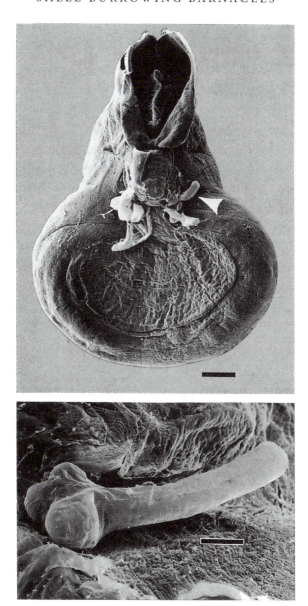

FIGURE 10.1. Top panel: A mature female burrowing barnacle, *Trypetesa lampas*, removed from her burrow. Only the exterior surface of her mantle is visible, and the opening to her mantle cavity can be seen at the top of the photograph. The arrow points to one of four dwarf males attached to her mantle. The scale bar indicates 0.5 mm. Bottom panel: A close-up of the male indicated by the arrow with his penis (not inflated) extending to the right. The scale bar indicates 0.1 mm. Photo reprinted with permission from Gotelli and Spivey (1992), figure 1.

forms, I had best begin by explaining what we see in the photographs in figure 10.1. The top photo shows a mature female that has been preserved and removed from her burrow by dissolving away the surrounding shell (in this case the shell of a Florida conch). We are viewing her as though seeing her through the shell wall to which she was attached. Her body is enclosed in a tough, sac-like mantle (the tissue that secretes the shelly plates of thoracican barnacles), and the orifice at the top of the picture is the opening of the cavity formed by the mantle. In the living animal the tissues of this opening would form bristly lips around the aperture in the host shell. Inside the mantle cavity (and not visible in the photo), the female's body is curled on itself so that both anterior (front) and posterior (back) ends are near the mantle opening. To picture this imagine yourself bent over and touching your toes inside a sleeping bag with your hands and feet at the open end of the bag. Instead of two feet and two hands the female barnacle has four pairs of modified legs called *cirri* that she uses like a net to trap food particles in the water and transfer these to her mouth. Three pairs of cirri are on the posterior part of the U-shaped body (akin to your legs and feet), and the fourth pair is on the anterior part (akin to your arms and hands). The female's mouth and the openings of her paired oviducts cluster near the bases of the anterior cirri (like your head between the bases of your arms). By contracting her mantle rhythmically and beating her cirri, the female sets up a current of water that brings food particles and oxygen in through the mantle opening and carries away carbon dioxide and wastes.

In the photograph we can see only the exterior surface of the mantle that would normally be pressed against the burrow wall, so you will have to be content with imagining the curled, shrimplike female within. Mature female *Trypetesa lampas* are large by burrowing barnacle standards, with mantle sacs averaging 5–11 mm in diameter. The large, round disk that makes up most of the visible mantle sac in the photograph is called the ovarian disk because the paired ovaries lie immediately beneath it, along the curved back of the female's body. It is flattened or disk-shaped to conform to the inner surface of the narrow burrow, which in turn must fit within the thickness of the host shell wall (usually about 2 mm).

Between the ovarian disk and the mantle opening you can just discern a small, raised circular knob of tissue. This is the horny stalk by which the female has cemented herself to the burrow wall. The white arrow to the right of this points to one of four males attached to the outer surface of the female's mantle, just below the attachment stalk, and the lower photograph shows this male in detail. His anatomy is straightforward. His long penis extends to the right and even though it is not inflated, it dwarfs his rounded, lobular body. To fertilize the female's eggs, the penis expands and stretches well into the mantle cavity so that the sperm can be deposited near the openings of the female's oviducts. Other than the penis the main constituents of the male's body are a single, large testis, a seminal vesicle for sperm storage and lipid globules stored for fuel. With no mouth, gut or digestive organs of any sort, it is clear that the males subsist only on their stored resources. Like *Osedax* males, their lifespan and sperm production must depend on the amount of fuel they have stored prior to maturation into adults.

The sexual dimorphism of adult *T. lampas* is truly extraordinary in every way. Females average ten times the length and 500 times the weight of their males,[10] and, once mature, the two sexes share virtually no morphological characteristics. If the adult males were not found living on the females and fertilizing their eggs, one would never identify them as members of the same species. Surprisingly, this remarkable sexual dimorphism develops only after the larval barnacles settle down and metamorphose to their juvenile forms. Both sexes go through a multistage larval development that is typical of most barnacles and similar to many other crustacean arthropods. They emerge from their mother's mantle cavity as free-swimming, triangular-shaped larvae, or *nauplii*, that look nothing at all like their future adult selves. Although the nauplii of many crustaceans actively feed, those of *T. lampas* and other burrowing barnacles do not. Like *Osedax* larvae, they depend entirely on the yolk provided by their mothers to fuel their development. They molt three times as nauplii, increasing in size only slightly with each molt, and then undergo their first metamorphosis, remaking themselves into another larval form, the *cyprid*.[11] The cyprid has a tiny bivalved shell and looks like a little swimming clam with eyes and legs. Its main job is to find a place to settle, and to that end it employs specialized appendages called antennules that stick out beyond the shell at the front of its body. Using its antennules, the cyprid walks over the surface of potentially suitable shells, sensing chemical cues, texture, and shell contours. On finding a suitable site it hooks on with its antennules and then anchors itself in place by excreting a mixture of proteins, phenols, and phenoloxidase that within a few hours has hardened into a rubbery, quinone-tanned protein. This insoluble cement forms the horny attachment stalk that fixes the cyprid to the substrate and ends its wandering life forever.

Up to this settlement stage males and females are indistinguishable, and their development is similar to that of other barnacles. However, within hours of settling, the cyprid larvae undergo yet another metamorphosis to become juveniles, and this is when the sexes diverge radically and when the peculiar traits of burrowing barnacles become apparent. If the larva is female it metamorphoses into a juvenile that closely resembles a miniature version of the adult form, bent double within its enveloping mantle. Initially the tiny female is attached to the outside of the host shell by the tips of her antennules, but she soon begins the process of burrow formation by excreting chemicals that dissolve the calcareous shell. As she grows through successive molts her mantle becomes covered with chitinous teeth that function like sandpaper. Contractions of the mantle scrape away at the shell, eventually excavating a long, thin burrow around her ever-growing body. The point of attachment to the shell accumulates a layer of cement with each molt until it becomes the thickened, horny knob evident in the photograph. The process of growing and creating a suitable burrow is long and slow, and it is six to nine months before the female matures her ovaries and begins producing batches of eggs. By then she has burrowed deeply into the shell and has expanded her mantle cavity and her burrow to accommodate broods of eggs. When her eggs are mature she releases them through her oviducts to be fertilized in the mantle cavity. The fertilized eggs remain within her mantle cavity until they hatch, and then the free-swimming nauplii exit through the mantle aperture to begin their own cycle of searching and settling.[12]

Of course, all of this can only happen if the female has succeeded in attracting at least one male to fertilize her eggs. Unlike larval spoon worms and bone-eating worms, larval *T. lampas* are genetically preset to be either male or female. This means that established females must attract male cyprids.[13] To increase the chances of this happening they entice passing males by releasing pheromones into the water. Male cyprids are attracted to these chemical cues and swim actively toward their source. Some females are very successful in attracting males and accumulate harems as large as seven to fifteen males. However, finding a mate is clearly a chancy business. When researchers Nicholas Gotelli and Henry Spivey surveyed the distribution of males on females in Apalachee Bay, Florida, they found that 45 percent of females had no males at all.[14] Far more females had no males than would have been expected if males were settling on females at random, and males were more likely to be found on larger and hence older females. Gotelli and Spivey postulated that the males settled passively on females in proportion to the surface area offered for settlement, resulting in an apparent preference for larger females. However, a second possibility is that larger females attract more males because they release larger volumes of pheromone. Still a third possibility is that the males actively select larger females because of their higher expected fecundity. Unfortunately these hypotheses have not been tested, so we do not know which if any of these possible explanations is correct. Nevertheless, Gotelli and Spivey's survey certainly suggests that failure to find a mate is a very real risk for female *T. lampas,* especially when they are young and small, and females may have to live for several seasons to achieve reproductive success. In fact they do live long lives for such small animals, typically surviving and continuing to grow for two to three years after settlement.[10]

This is the story for female *T. lampas,* but if the cyprid larva is male, an entirely different fate awaits. During metamorphosis to the juvenile stage, males resorb (digest) most of their organ systems and emerge slightly smaller than they were as cyprids. The stripped-down juvenile males then develop their reproductive systems at a frenzied pace. Within twenty-four hours of settlement the large testis has clearly differentiated, and within a few days the tiny males are fully mature and capable of fertilizing eggs.[15]

This rapid maturation is only helpful if the precocious males have actually settled on females, and so, as cyprids, the males spend considerable time searching for a suitable mature female. Using cues detected by their antennules they seem to be capable of selecting particular shells and then choosing particular females within that shell. Once they have zeroed in on their female of choice, they explore and probe her mantle, seeking the best position to attach themselves. They prefer to settle on the forward part of her ovarian disk near the horny stalk that attaches her to the shell and close to the mantle opening, where the four males are clustered in the photograph (figure 10.1). However, since females often host more than one male (35 percent of females in Gotelli and Spivey's survey), new arrivals may find that other males have settled before them and taken the best spots. Latecomers avoid settling close to entrenched rivals and so tend to settle further away from the mantle aperture than they would in the absence of competition. Since each male is cemented in place, the newcomers cannot oust established males by any sort of physical intimidation as if they were tiny elephant seals. Instead, the competition for position must favor males that arrive early enough to grab the best settlement sites and males that possess a very long and extensible penis, so that they can fertilize females even if they fail to acquire a front-row position. The presence of more than one male on a given female also means that, once settled, the males must compete to fertilize her eggs. It is likely that the advantage in this struggle accrues to males that produce copious amounts of sperm and are able to rapidly replenish their sperm supplies for multiple mating attempts. Seen in this light it is not surprising that the males devote most of their body mass to their large testis at the expense of their somatic organ systems. The tiny males may also need an extra-large testis and a super-long penis, relative to their own body size, just to compensate for the size differential between the sexes. After all, it is a long way from even the most advantageous settlement site to the female's oviducts, and copious amounts of sperm are likely needed to fertilize her entire batch of eggs.

Although competition among males for the best sites on females should favor males that are ready to settle early, there is no evidence that males complete their larval development faster than females. As far as we know the two sexes go through the various larval stages at the same rate. They differ enormously in age at maturity, but this is entirely caused by the extremely rapid maturity of males after the cyprid larva has settled and molted. Only after settlement, from the juvenile stage onward, do the two sexes run on different timescales. The males forgo growth, mature quickly, and die young, spending about 97 percent of their post-settlement life as fully reproductive, mature adults. Most live only three to four months, about a tenth as long as their females, and will have matured, mated, and died by the time the females in their cohort have even become sexually mature. (The mates they seek must be females from previous cohorts that have had sufficient time to settle and mature.) Males are more likely to survive to reproductive age than females simply because they do not have to survive for so long before becoming mature, but they likely live through only one breeding season. In contrast the females grow throughout their post-settlement lives, take time to grow before maturity, live through several breeding seasons, and spend only about 75 percent of their post-settlement lives as fully mature adults. Whereas males can mate with only one female, the females often accumulate more than one male at a time and recruit new males each breeding season. This sexual divergence in life histories is equally as impressive as the morphological differences between the sexes.

The parallels between *T. lampas* and the *Osedax* from chapter 9 are both striking and informative. Both have long-lived, sessile females and short-lived, dwarf males that live attached to their mates. In both the larvae disperse and settle without feeding, living solely on the nutrients supplied by their mothers in the egg. In both, the males rapidly mature after settlement into structurally and functionally simplified adults. They never feed and live very short adult lives, constrained by their energy reserves. Females, in contrast, mature into complex, fully functional adults that feed and grow before maturity and have relatively long adult lives through which they spawn repeatedly and accumulate multiple mates. In both bone-eating worms and burrowing barnacles the differences between the sexes emerge only after larval settlement and so must be beneficial only during the adult, sessile phase of the life cycle. This means that the benefits of small size must accrue to males only after they have found a mate. This is somewhat surprising given what we have seen in our other examples of species with dwarf males. In orb-web spiders, blanket octopuses, and seadevils, the size differences between the sexes develop before males begin searching for mates, and small size has been assumed to benefit males in some way during their search for widely dispersed females.[16] Bone-eating worms and burrowing barnacles demonstrate definitively that this need not be the case. Dispersal and mate search are important components of the life histories of both *Osedax* and *T. lampas*, and yet males and females do not diverge in size until after settlement. Put simply, males do not become dwarfs until after they have found a mate, and so dwarfism cannot be an adaptation for finding mates.

If energy efficient mate searching is not the primary reason why males become dwarfs, than what is? One characteristic that all our example species have in common is that the dwarf males become sexually mature much faster than their females, and this is generally true of all species with dwarf males across the animal kingdom. Could it be that the main reason why dwarf males are so small is simply that this enables them to forgo growth and mature at a very young age? The differences between dwarf males and their females in age at first reproduction can be so extreme that the males in many species will live their whole lives before the females in their cohort have even begun to reproduce. Early maturation has three key benefits for males. It increases the probability that they will survive to reproductive age; it increases the probability that they will be reproductively competent when and if they are fortunate enough to encounter a female; and in seasonal breeders it increases the chances of mating with a female before she has mated with other males. In the last case even if first-arriving males cannot fully monopolize matings with their females, they may be able to obtain the most favorable mating positions (as in *T. lampas*) or in some other way impede the fertilization success of later-arriving males (as in *Argiope* males leaving behind their broken embolus caps and dying *in copula*). Perhaps we need look no further to explain the small size of dwarf males than to conclude that they are small because they choose early maturation over growth, and it is the former that is key to their Darwinian fitness.[17]

Given what we have discovered about the animals in this book, I tend to think that this explanation of male dwarfism is at least partly correct. The main driving force behind the evolution of dwarf males is probably selection for early maturation rather than direct selection for small size. However, this does not preclude other, more direct benefits of very small size in males. For example, in burrowing barnacles and bone-eating worms males obviously have to be small enough to fit into the cramped quarters of the female's burrow or tube. Given that these males do not eat at all throughout their lives, being small also allows them to live longer and produce more sperm using the finite resources they carry with them from the egg. (Why they do not eat is another question, the answer to which probably has to do with living in very resource-poor environments.[18]) In seadevils the males become parasites on their females, and this surely limits how large they can be because resources diverted to their maintenance decrease the resources that the female can allocate to herself or to producing eggs. In blanket octopuses males do forage, if somewhat inefficiently, while searching for females, and their small size probably allows them to survive longer during this phase of their lives. Being small also allows them to disperse efficiently by hitching rides on the bells of floating jellies and to protect themselves with stinging tentacles purloined from man o' war encountered along the way. Finally, the small size of male orb-web spiders probably enables them to move more efficiently through tangled vegetation in search of a mate. In all of these species I suspect that the males are small not only because this is an indirect consequence of selection favoring early maturation, but also because small size itself is beneficial in some way. It may well be that a combination of both types of advantages is necessary to prompt the evolution of truly dwarf males.

CHAPTER 11

The Diversity of Sexual Differences

Differences between Males and Females
across the Animal Kingdom

The examples we have seen, from elephant seals to bone-eating worms, are among the most extreme sexual dimorphisms found in the animal kingdom. As promised, these extraordinary species illustrate the truly amazing diversity of ways in which animals can divide their reproductive roles into male and female functions. However, the extreme differences between sexes in these species are obviously not typical of animals in general. The males and females in most animal species do not differ so starkly. In this chapter I shift gears to ask in what ways male and female animals tend to differ on average. In other words, "What are the more typical patterns of sexual dimorphism in the animal kingdom?" To answer this question I cataloged how males and females have been distinguished within each of the twenty-six phyla that contain dioecious species. To increase the resolution of my survey, I cataloged the sexual differences separately for each taxomomic class within each of these phyla, for a total of seventy-three classes.[1] I restricted my catalogue to externally apparent morphological traits because these are the only types of traits recorded for many of the lesser known animal taxa. Even in better-known taxa we seldom know enough about the physiology, ecology, life history, or behavior of the animals to make meaningful comparisons between the sexes. However, a bewildering diversity of sexually dimorphic traits have been described, and these provide plenty of fodder for discerning general patterns. For simplicity I organized the various morphological traits into seven broad categories. Two of these, externally visible gonads (testes or ovaries) and genital openings (often called *gonopores*), are components of reproductive tracts and so, by definition, are primary sexual dimorphisms. The remaining categories are all secondary sexual dimorphisms and include body size, body shape, appendages (number, size, or shape), integumental morphology (characteristics of the external body covering other than color), and color (used loosely to include all pigmentation and structural color).

To determine if the patterns of sexual dimorphism differ among the major evolutionary lineages of animals, I summarized the patterns separately for four different phyletic clusters (table 11.1). Three of these are distinct evolutionary lineages of animals that share fundamental aspects of their ontogenetic development and basic body plans. The first, the Deuterostomia, is dominated by our own phylum Chordata (primarily vertebrates), whereas the second, the Ecdysozoa, is dominated by the huge phylum Arthropoda (animals with hard, external skeletons and jointed legs such as insects, spiders, and crustaceans). The third lineage, with the tongue-twisting name Lophotrochozoa, is dominated by shelly and wormy animals including mollusks (Mollusca), flatworms (Platyhelminthes), and segmented worms (Annelida). Of the seventy-three classes containing dioecious species, sixty-five fall into one of these three major lineages. The excluded classes are distributed across only three phyla: the sponges (Porifera), jellies (Cnidaria), and comb jellies (Ctenophora). Although these three phyla are not particularly closely allied with each other, for simplicity I treat them as a fourth cluster, the "nonbilaterian phlya," so named because many zoologists refer to the Deuterostomia, Ecdysozoa, and Lophotrocozoa collectively as the Bilateria.

Below I summarize the broad trends and patterns of sexual dimorphism for all classes combined, as well as separately, within each of the four phyletic clusters. The proportions and percentages given refer to relative frequencies only among the seventy-three classes that contain at least some dioecious species, and the question addressed is thus: "Where separate sexes occur, how frequent are the various forms of sexual dimorphism?"[2]

TABLE 11.1.

Phyla and classes included within each of the four major clusters
used in the survey of sexually dimorphic traits

Phyla	*Classes*
Deuterostomia	
Chordata	Actinopterygii, Amphibia, Aves, Cephalaspidomorphi, Cephalochordata, Elasmobranchii, Holocephali, Mammalia, Reptilia, Sarcopterygii
Echinodermata	Asteroidea, Crinoidea. Echinoidea, Holothuroidea, Ophiuroidea
Hemichordata	Enteropneusta, Pterobranchia
Ecdysozoa	
Arthropoda	Arachnida, Branchiopoda, Chilopoda, Diplopoda, Entognatha, Insecta, Malacostraca, Maxillopoda, Merostomata, Ostracoda, Pauropoda, Pycnogonida, Symphyla
Cephalorhyncha	Kinorhyncha, Loricifera, Priapulida
Nemata	Adenophorea, Secernentea
Nematomorpha	
Onychophora	
Tardigrada	Eutardigrada, Heterotardigrada
Lophotrochozoa	
Acanthocephala	Archiacanthocephala, Eocanthacephala, Palaeacanthocephala
Annelida	Polychaeta
Brachiopoda	Articulata, Inarticulata
Cycliophora	Eucycliophora
Echiura	Echiurida
Ectoprocta	Gymnolaemata
Entoprocta	
Mesozoa	Orthonectida
Mollusca	Aplacophora, Bivalvia (Pelecypoda), Cephalopoda. Gastropoda, Monoplacophora, Polyplacophora, Scaphopoda
Nemertea	Anopla, Enopla
(Rhynchocoela)	
Phoronida	
Platyhelminthes	Trematoda, Turbellaria
Rotifera	Eurotatoria, Pararotatoria
Sipuncula	
Nonbilaterian Phyla	
Cnidaria	Anthozoa, Cubozoa, Hydrozoa, Scyphozoa, Staurozoa
Ctenophora	Tentaculata
Porifera	Calcarea, Demospongia

NOTE: Taxa are listed in alphabetical order within each cluster, and only those containing at least some dioecious species are shown.

SOURCES: Taxonomic classification from Adoutte et al. (2000), Hickman et al. (2007), and Maddison and Schulz (2007).

Primary Sexual Dimorphisms

Other than the genital region, components of the reproductive tract are seldom visible externally. However, some species are sufficiently transparent that the gonads can be seen through the body wall, at least during the active breeding season. If this is the case a keen observer can usually distinguish ovaries from testes by size, texture, shape, or (most often) color. These visible gonadal dimorphisms are common in the transparent jellies (Cnidaria) and comb jellies (Ctenophora), but they are otherwise uncommon, occurring in only 23 percent of the seventy-three classes distributed across seven phyla.[3]

As might be expected, sexual differences in genital openings are much more prevalent, having been described in 53 percent of classes across fifteen different phyla. They are most common in the Ecdysozoa and Deuterostomia (73 percent and 71 percent), somewhat less common in the Lophotrochozoa (42 percent), and absent from the nonbilaterian phyla. The differences among the four phyletic clusters can largely be attributed to differences in the prevalence of copulation in these groups: copulation of some form has been reported for 63 percent of classes in the Ecdysozoa, 59 percent of classes in the Deuterostomia, 44 percent in the Lophotrochozoa, and not at all in the nonbilaterian phyla. In species that copulate the genitalia of both sexes are usually modified to accommodate internal fertilization by means of male genitalic intromittent organs.[4] Less commonly female genitalia are modified to accept nongenitalic intromittent organs, such as palps in spiders, or for the attachment of spermatophores produced by males, as in centipedes and many insects. Rarely, genital dimorphisms are found in a species that simply release sperm into the water, and the functional significance of these odd sexual features remains a mystery.[5]

If we consider gonadal and genital dimorphisms combined, we find externally apparent primary sexual dimorphisms in 70 percent of the classes that contain dioecious species, spread across nineteen phyla (figure 11.1). The prevalence is somewhat lower in the Lophotrochozoa (58 percent) than in the Deuterostomia, Ecdysozoa, and nonbilaterian phyla (73–76 percent), probably because lophotrochozoan species tend to be both less

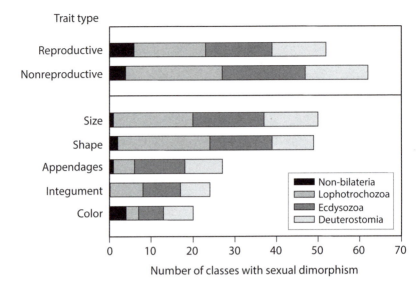

FIGURE 11.1. The number of animal classes for which externally apparent sexual dimorphisms have been reported in characteristics of the reproductive tract (primary sexual dimorphisms) and in nonreproductive traits (secondary sexual dimorphisms). The latter category is broken down into body size, body shape, appendages, integumental morphology, and color. The stacked bars represent the numbers of classes within each of the four major phyletic clusters described in the text.

transparent than nonbilaterian species (so their gonads are less likely to be visible) and less likely to copulate than deuterostomes or ecdysozoans (so less likely to have genital dimorphisms).

Secondary Sexual Dimorphisms

Secondary sexual dimorphisms are considerably more common than externally visible primary sexual dimorphisms, occurring in 84 percent of the classes across twenty-three phyla (figure 11.1). Animals in the bilaterian lineages are more likely to show secondary than primary sexual differences, with prevalence ranging from 91 percent in the Ecdysozoa to

85 percent in the Lophotrochozoa. In contrast, secondary sexual dimorphisms are much less common in nonbilaterian species, occurring in only 50 percent of classes, and nonbilaterian species are more likely to show primary than secondary sexual differences.

Why secondary sexual dimorphisms are so much less common in nonbilaterian animals is an intriguing question. Unfortunately our knowledge of the various components of Darwinian fitness in these animals is too limited to provide the answer. However it may be telling that dioecy itself is relatively uncommon in these taxa (appendix B) and that most species spawn their eggs and sperm into the water with no obvious interactions between the sexes. These observations indicate little specialization for separate sex roles in the nonbilaterian phyla beyond the very basic allocation of testes and ovaries to separate individuals in a minority of species. Why this should be so remains a mystery, but it is clear that pronounced sexual specialization is largely restricted to bilaterian lineages.

Sexual Dimorphisms in Body Size

Of the five categories of secondary sexual dimorphisms, differences in overall body size are the most frequently reported. Size dimorphisms have been reported in 67 percent of the classes that contain dioecious species, distributed across twenty phyla. They are prevalent in the Bilateria, with proportions ranging from 69 percent in the Lophotrocozoa to 77 percent in the Ecdysozoa, but among nonbilaterian phlya they have been noted in only a few sea anemones (Cnidaria, Anthozoa) (figure 11.1). The prevalence of sexual size dimorphism among bilaterian animals is not unexpected given the close correlations between body size and so many other aspects of animal biology in these lineages. As we noted in chapter 1, marked differences between the sexes in traits such as body shape, life history, behavior, and ecology are almost always accompanied by differences in size, and vice versa. Chapters 3 through 10 illustrated these relationships in the extreme, but the same principles hold for the less extreme size dimorphisms typical of most bilaterian animals.

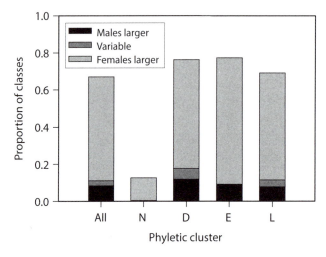

FIGURE 11.2. The proportion of the animal classes containing dioecious species ($n = 73$) that have sexual differences in average body size. The stacked bars indicate the proportion of classes in which males average larger than females (black); the sexes differ in size but which sex is larger varies among species, and there is no overall trend for males or females to be the larger sex (dark gray); or females average larger than males (light gray). Proportions are shown for all classes combined (left bar) and for each of the major phyletic clusters separately. N, nonbilaterian phyla; D, Deuterostomia; E, Ecdysozoa; L, Lophotrochozoa.

Given that male and female animals often differ in size, the obvious question is "Which sex is larger?" From our human perspective we might presume that males are generally larger, and indeed this is the case for the majority of mammals and birds. However, this generality quickly becomes invalid as we move to less familiar animal groups (figure 11.2). Females average larger than males in all the vertebrate classes other than mammals and birds,[6] in the vast majority of arthropods, including insects and spiders,[7] and in almost all the "wormy" phyla and classes.[8] Sexual size dimorphism is not as prevalent in other phyla, but when it has been noted, females are again almost always the larger sex. This is true for most hard-shelled animals (mollusks and brachiopods), most echinoderms (anemones, sea cucumbers, sea lilies, and sea stars), various phyla of small, free-living animals including water bears and rotifers, and the vast majority of parasitic and commensal species (i.e., species that live in or on other animals).[9] Overall, female-larger dimorphism predominates in 86 percent of classes that have some form of sexual size dimorphism, a trend that holds over all four phyletic clusters. Clearly, when the sexes differ in size, females are more likely to be the larger sex.

Most species have much more modest size dimorphisms than those I have emphasized in previous chapters, with the sexes typically differing in size by no more than 50 percent in linear dimensions and more typically by only 5 percent to10 percent. However, in almost all of the phyla and classes in which sexual size dimorphism has been quantified, species with extreme sexual size dimorphisms have been discovered (table 11.2). These quantitative data reveal enormous variation in the magnitude of size dimorphism within individual classes or even within individual orders (taxonomic subdivisions of classes). In every phylum, class, or order listed in table 11.2, the range of recorded size dimorphisms extends from species in which males are the larger sex to species in which females are the larger sex. By far the greatest range of dimorphisms occurs in the ray-finned fishes, with giant seadevils at one extreme and shell-carrying cichlids at the other. In mammals the overall range is less impressive, but almost the full range occurs within a single family, the eared seals (family Phocidae). This family includes both southern elephant seals, the most size dimorphic of all mammal species with males up to eight times heavier than females, and Leopard seals, where females average as much as 64 percent heavier than males.[10] Among birds the greatest range of dimorphisms occurs within a single order, the perching birds (order Passeriformes), which includes the species with the greatest female-larger size dimorphism (the dappled mountain robin, with females 2.17 times heavier than males) and the species with the third greatest male-larger dimorphism (the brown songlark, with males 2.31 times heavier than females).[11] These huge ranges of sexual size dimorphism within groups of quite closely-related species tell us that, although each animal clade does have an average pattern and magnitude of dimorphism (such as moderate male-larger dimorphisms in mammals), individual species within that clade are clearly not constrained to that pattern.

TABLE 11.2.

Quantitative estimates of sexual dimorphism for body size in various animal phyla and classes

Phylum	Class (Order)	Common name	Larger sex[a]	Mass or volume SD			Linear SD			Sources
				Mean	Min	Max	Mean	Min	Max	
Acanthocephala		Spiny-headed worms	f	4.0	-1.7	62				1
Arthropoda	Arachnida	Spiders	f				1.3	-1.1	4.5	2
	Branchiopoda (Diplostraca)	Water fleas, cladocerans	f				2.0	-1.1	3.0	3
	Insecta	Insects	f		-1.2	2.3		-1.1	2.0	4
	Maxillopoda (Calanoida)	Calanoid copepods	f				1.1	-1.4	2.1	3
	Maxillopoda (Cyclopoida)	Cyclopoid copepods	f				1.4	-1.4	3.4	3
Chordata	Actinopterygii	Ray-finned fishes	f		-13	5×10^5		-2.5	10	5
	Amphibia	Amphibians	f			40		-1.7	3.1	6
	Aves	Birds	m	-1.04	-3.1	2.2	-1.03	-1.7	1.3	7
	Elasmobranchii	Sharks	f				1.1	-1.4	1.6	8
	Reptilia (Squamata)	Lizards and snakes	m			13		-1.5	2.1	9, 10
	Reptilia (Testudines)	Turtles	f				1.1	-1.4	2.8	9, 11
	Mammalia	Mammals	m	-1.2	-7.8	1.7		-1.6	1.2	12
Nemata[b]		Roundworms	f				1.5	-1.3	3.0	13

NOTE: Sexual dimorphism (SD) is given as the mean size of the larger sex divided by the mean size of the smaller sex. Because females are the larger sex in most groups, male-larger dimorphisms are indicated as negative (i.e., with a minus sign). Ratios less than 10 have been rounded to one decimal place except for the mean values for birds, where two decimal places are retained for clarity.

[a]Sex that is larger in the majority of species.

[b]Parasitic species only.

SOURCES: 1, Poulin and Morand (2000); 2, Foellmer and Moya-Laraño (2007); 3, Gilbert and Williamson (1983); 4, Fairbairn (1997), Blanckenhorn et al. (2007); 5, Clarke (1983), Schultz and Taborsky (2000), Pietsch (2009); 6, Kupfer (2007) and personal communication, Kraus (2008); 7, Székely et al. (2007); 8, Sims (2005); 9, Cox et al. (2007); 10, Pearson et al. (2002); 11, Gibbons and Lovich (1990); 12, Lindenfors et al. (2007), Alexander et al. (1979), Ralls (1976); 13, Poulin (1997).

Why sexual size dimorphism varies so greatly among and within animal taxa has been the subject of a large body of research and considerable speculation.[12] Although we are far from having the complete answer, evidence from many taxa of vertebrates and arthropods supports the hypothesis that one major driver is variation in the strength and direction of sexual selection on body size in males. As we saw in chapters 3 through 5, sexual selection on males is particularly strong in polygynous species where the successful competitors are able to monopolize mating opportunities with many females, and if the competing males engage in physical contests, larger males have the advantage.[13] This form of sexual selection is most extreme in species such as elephant seals and shell-carrying cichlids where breeding females tend to group together because resources such as food or birthing sites are limited and patchily distributed.[14] The aggregation of females in these species enables males to more easily defend whole groups of females from mating attempts by rival males. In birds polygyny and its attendant sexual selection typically derives from a rather different mating system. Competition among male birds for mating opportunities tends to be most intense in species where males gather in a restricted display area (a lek) and attempt to attract mates through visual and auditory displays. As we saw for great bustards, the males in such species compete with each other mainly through their displays, but they also physically chase and attack one another as they jostle for dominance or for prime display sites. Larger males tend to be more successful in these aggressive interactions with other males, especially if the mating competitions take place on a solid substrate.[15] In both types of polygynous mating systems females typically augment the large male advantage by preferring to mate with the males that are most successful in male-male interactions and by rebuffing attempts by other males attempting to sneak copulations while the dominant male is otherwise occupied.

Sexual selection favoring large males clearly promotes the evolution of male-larger size dimorphisms, not only in mammals, birds, and fishes, but also in many species of reptiles, amphibians, insects, spiders, and malacostracan crustaceans (lobsters, crabs, and shrimp).[16] However it is doubtful that sexual selection on males is sufficient on its own. Just as critical in determining the ratio of male to female size is the pattern of selection acting on body size in females. Large size can convey a number of advantages to females. They may produce larger and better-nourished offspring, provide superior maternal care, or even be favored by sexual selection.[17] In most taxa, however, the main factor favoring large female size is a strong positive relationship between body size and fecundity. The influence of fecundity on Darwinian fitness is so strong in females that if fecundity scales strongly with body size, females are likely to be the larger sex even if sexual selection favors large size in males.[18] Given this, it seems likely that the evolution of pronounced male-larger size dimorphism requires not only strong sexual selection favoring large size in males but also relaxation of the relationship between fecundity and body size in females, as we saw in the examples of great bustards, elephant seals, and shell-carrying cichlids.

In most animal groups the balance of selection on males and females plays out so that females, not males, are the larger sex[19], and in many species they are very much larger than their mates. This is clearly evident in table 11.3, which shows examples of species in which one sex is more than twice as large as the other in linear measures or at least eight times as heavy, the standard criterion for defining males as dwarfs relative to their mates. In assembling these data I found species with dwarf males in twenty-three classes across twelve different phyla and in all three bilaterian lineages. I found many more examples than I had room to list and had to cull these down to only a few examples for my table. However, when I applied the same criterion to identify species with the reverse dimorphism (dwarf females/giant males), only one species qualified: the shell-carrying cichlid, *Lamprologus callipterus* from chapter 5. No mammals or birds made the cut-off, in spite of the prevalence of male-larger dimorphisms in those classes. The most extreme sexual size dimorphisms are found exclusively in species where females are the larger sex.

We still do not have a firm, general explanation for the occurrence of male dwarfism in so many species. The most widely proposed hypothesis is that wide dispersal of individuals at very low densities across low resource environments favors the evolution of large, long-lived females and small, early-maturing males.[20] According to this hypothesis high juvenile mortality rates combined with very low encounter rates between mature males and females favor males that mature early and are specialized for energetically efficient, long distance dispersal, mate location, and sperm production. Many of these males spend a high proportion of their lives simply searching for a mate, and those that succeed in finding one typically remain with her, often as parasites or commensals. Conversely, in females, high juvenile mortality combined with low encounter rates between sexes has favored the evolution of large size, high fecundity, and longevity because the females have to grow large in order to accommodate large numbers of eggs, and they have to live long enough to accumulate nutrients for those eggs and mates to fertilize them. According to this general scenario the disparate selection pressures on the two sexes have resulted in very small, early-maturing males combined with very much larger and later-maturing females. Although the species we met in chapters 6 through 10 certainly fit this general pattern, biologists have not yet been able to test all the components of the hypothesis, and so we cannot say for sure which, if any, are both necessary and sufficient for the evolution of these truly extreme sexual dimorphisms.

Sexual Dimorphisms in Body Shape

Differences in body shape are the second most prevalent category of secondary sexual dimorphisms, having been described in 66 percent of animal classes that contain dioecious species, including 59 percent of deuterostomes, 68 percent of ecdysozoans, 81 percent of lophotrochozoans, and 25 percent of nonbilaterian classes (figure 11.1). In the lophotrochozoans and nonbilaterians, shape dimorphisms are even more common than size dimorphisms. The most common shape dimorphism is for females to have thicker bodies than males. This is common in all types of animals, including active, streamlined swimmers such as sharks, bony fishes, and lampreys; slow-moving terrestrial animals such as millipedes, snails, and slugs; sessile animals such as sea anemones; and various "wormy" animals including segmented worms (Annelida), roundworms (Nemata), and horsehair worms (Nematomorpha).[21] It seems likely that

TABLE 11.3.

Examples of animal species from different classes illustrating extreme sexual dimorphism for body size

Phylum	Class	Common name or general description	Scientific name	Larger sex	Size ratio (larger/smaller) Mass or volume	Size ratio (larger/smaller) Linear	Source
Acanthocephala	Archiacanthocephala	Spiny-headed worm; parasite of birds	*Mediorhynchus mattei*	f	15	2.8	1, 2
	Eoacanthocephala	Spiny-headed worm; parasite of fish	*Acanthosentis dattai*	f	9.8		2
	Palaeacanthocephala	Spiny-headed worm; parasite of fish	*Hemirhadinorhynchus leuciscus*	f	62		2
Annelida	Polychaeta	Bone-eating marine worm	*Osedax rubiplumus*	f		57	3
		Interstitial marine worm	*Dinophilus gyrociliatus*	f		30	4
Arthropoda	Arachnida	Giant golden orb-weaver	*Nephila turneri*	f		9.6	5
		Yellow garden spider	*Argiope aurantia*	f		3.5	6
	Branchiopoda	Cladoceran (water flea)	*Daphnia cephalata*	f	53	2–6	7
	Insecta	Tessellated stick insect	*Acrophylla tessellata*	f		2.0	8
	Maxillopoda	Burrowing barnacle	*Trypetesa lampas*	f	500	7.8	9
		Copepod; parasite of sharks	*Kroyeria caseyi*	f		33	10
Chordata	Malacostraca	Isopod; parasite of prawns	*Boptrus manhattensis*	f	56	4	11
	Actinopterygii	Giant seadevil	*Ceratias holboelli*	f	500,000	60	12
		Deep-sea stalkeye fish	*Idiacanthus fasciola*	f	50–100	5–10	13
		Shell-brooding cichlid	*Lamprologus callipterus*	m	13	2.4	14
	Amphibia	Wrinkled ground frog	*Platymantis boulengeri*	f	40	3.1	15
	Reptilia	Australian carpet python	*Morelia spilota*	f	13	2.1	16

Cycliophora	Eucycliophora	Cycliophoran; commensal on lobsters	*Symbion pandora*	f		11	17
Echinodermata	Ophiuroidea	Brittle star; commensal on sand dollars	*Ophiodaphne formata*	f		5	18
Echiura	Echiurida	Green spoon worm	*Bonellia viridis*	f		23–70	4, 19
Mollusca	Bivalvia	Clam; commensal on sea cucumbers	*Montacuta percompressa*	f		10	20
	Cephalopoda	Blanket octopus	*Tremoctopus violaceous*	f	10,000–40,000	100–150	21
	Gastropoda	Limpet; parasite on sea stars	*Thyca crystallina*	f		10	22
Nemata	Adenophorea	Bladder thread worm; parasite of rats	*Trichosomoides crassicauda*	f		4.6	23
	Secernentia	Roundworm; parasite of fish	*Camallanus xenentodoni*	f		2.6	24
Nemertea	Enopla	Ribbon worm; parasite of crabs	*Carcinonemertes pinnotheridophila*	f		3.7	25
Platyhelminthes	Turbellaria	Flatworm; parasite of amphipods	*Kronborgia amphipodicola*	f		5.6	23
Rotifera	Eurotatoria	Monogont rotifer	*Brachionus plicatilis*	f	6.3	2	26

NOTE: Sexual size dimorphism is defined as extreme if the ratio of body sizes (larger sex/smaller sex) is at least two for linear dimensions or at least eight for mass or volume.

SOURCES: 1, Marchand and Vassilliades (1982); 2, Poulin and Morand (2000); 3, Rouse et al. (2004); 4, Giese and Pearce (1975b); 5, Kuntner and Coddington (2009); 6, Matthias Foellmer, personal communication; 7, Hebert (1977); 8, Sivinski (1978); 9, Gotelli and Spivey (1992); 10, Benz and Deets (1986); 11, Gissler (1882); 12, Berterlsen (1951); 13, Clarke (1983); 14, Schultz and Taborsky (2000); 15, Kraus (2008); 16, Pearson et al. (2002); 17, Kristensen (2002), Obst and Funch (2003); 18, Tominaga et al. (2004); 19, Greenwood and Adams (1987); 20, Chanley and Chanley (1970); 21, Norman et al. (2002); 22, Elder (1979); 23, Adiyodi and Adiyodi (1992); 24, Poulin (1997); 25, McDermott and Gibson (1993); 26, Epp and Lewis (1979).

this general female thickness has to do with the production and storage of eggs, and this is clearly the case in species where females have enlarged brood pouches or abdomens. However, it is also possible that the thinner body morphology of males facilitates efficient mate searching. The second most common shape dimorphism is for male body shapes to be adapted for grasping and holding females during mating. The curved tails of male nematode worms, bent abdomens of some insects, and concave plastrons (bottom shells) of male turtles are examples of this. Male body shape may also be adapted for success in physical contests with other males. For example, in species where males butt heads or use horns or antlers in contests with other males, they often have proportionally larger heads and upper bodies than females of the same body size. Sheep, deer, horned beetles, and stalk-eyed flies are good examples of this.[22]

Although these examples of shape dimorphisms can be readily explained as adaptations for male and female reproductive roles, in other cases the functional significance of the shape dimorphisms remains obscure. For example, zoologists have yet to fully understand why female snakes and lizards tend to have relatively longer bodies but smaller heads and shorter tails than males, although fecundity selection favoring large abdomens in females combined with sexual selection favoring large heads in males does seem the most likely explanation.[23] Shape dimorphisms are even more enigmatic in several of the less well-studied animal groups, such as water bears and centipedes, and these will probably remain so until we learn more about the basic biology and behavior of these animals.

Sexual Dimorphisms in Appendages

In addition to differences in overall body size and shape, males and females often differ in the size or shape of their appendages, or one sex may have a unique appendage not present in the other sex. Appendage dimorphisms are considerably less common than size and shape dimorphisms (figure 11.1), mainly because many animals do not have appendages. I once asked my son (who has since gone on to become a computer engineer, not a biologist) what he recalled most about animal diversity from his introductory college course in biology. His reply was that his main impression was that most animals are either arthropods or worms. This is certainly an apt appraisal of animal diversity given that there are fifteen classes of arthropods, and animals called worms of one form or another are found in fifteen phyla and twenty-three classes. Given the popularity of the worm phenotype, not to mention other types of legless animals such as clams, sea cucumbers, and jellies, it is not surprising that we find appendage dimorphisms in only 37 percent of the classes that have dioecious species. For the same reason it is also not surprising that appendage dimorphisms are far more common in the deuterostomes and ecdysozoans (53 percent and 55 percent of classes respectively), than in the lophotrochozoans (19 percent) and nonbilaterians (13 percent). Appropriately, appendage dimorphisms are most prevalent in the phylum that specializes in appendages, the Arthropoda (whose name comes from the Greek for "jointed feet"). Many male arthropods have antennae or legs that are modified for grasping females, for sexual signaling, or to act as armaments in contests with other males. These modified appendages are typically larger and more robust than the female counterparts and may have more complex shapes and extra bristles, hooks, or even hinges. Arthropod males have also co-opted various feeding or locomotory appendages to serve as copulatory organs or for passing spermatophores.[24] In insects, wings are often modified to send visual or acoustic sexual signals, and antennae are often modified to detect air-borne chemical signals (pheromones) from potential mates. In some insects one sex may be winged and the other not, or the wing size may differ between the sexes, dimorphisms that presumably reflect different allocation to flight versus reproduction in the two sexes.[25] In a few arthropod classes appendages have even been modified for brooding eggs. For example, many females in the class Malacostraca (crabs, lobsters, and shrimp) use modified swimming appendages (called pleopods) or specialized plates called oostigites for brooding eggs, whereas in another arthropod class, the sea spiders (Pycnogonida), males have specialized legs called ovigera for this task.

Male appendages are also modified in phyla other than the Arthropoda, most commonly for grasping females during mating. For example, many male reptiles and amphibians have front limbs modified for grasping females; sharks, rays, chimeras, and ratfishes have pelvic fins modified as sexual claspers; and even tiny male water bears have longer legs than females, presumably because this helps them clasp females during copulation. Male appendages have also been modified to act as copulatory organs in several classes. The modified pelvic fins (called gonopodia) in some ray-finned fishes and hectocotyl arms in squid and octopods are familiar examples. Less commonly appendage dimorphisms reflect male adaptations for attracting mates (e.g., the elaborate fins of some ray-finned fishes) or for success in male-male combat (e.g., long legs in roosters and peacocks).

Although there is little morphological homology between the appendage modifications in different animal classes, the parallels in function are obvious. Modifications that enable males to grasp females during mating are the most common, but modifications for sperm transfer, sexual signaling, sexual combat, and even egg brooding have also evolved independently in different animal classes. Because these dimorphisms are often discrete (i.e., one type of appendage occurs in one sex and is absent or has a very different size or shape in the other), they are often useful as distinguishing features of the sexes, particularly in species with only moderate dimorphisms in size or shape.

Sexual Dimorphisms in Integumental Morphology

Dimorphisms in integumental morphology include all dimorphisms in external body covering other than those in the area closely surrounding the genital openings, which I classified as genital dimorphisms, and color, which has a separate category of its own. Teeth and claws are included as integumental traits because they are derived from the same cell layer that gives rise to the external body covering. Sexual dimorphisms in integumental traits are even less common than appendage dimorphisms, having been noted in only 33 percent of classes with dioecious species (41 percent of deuterostomes, 41 percent of ecdysozoans, 31 percent of lophotrochozoans, and no nonbilaterian classes) (figure 11.1). Often, males have modified spines, scales, folds, or plates that function as mini-appendages for grasping females. The clavoscalids of lorciferans (in the phylum Cephalo-rhyncha), the long claws of male water bears, and the longitudinal fold on the heads of some male rotifers are examples of these types of dimorphisms. Males may also have specialized spines that serve as nongenital copulatory organs, as in the kinorhynchans (another class in the phylum Cephalo-rhyncha) and roundworms. In many vertebrate species teeth, spines, horns, or antlers serve as sexual signals and as weapons in male-male contest com-petition. Dimorphisms in the distribution, length, or density of hairs or feathers may also serve as sexual signals. In arthropods males often have pegs, spines, bristles, or horns that serve as weapons in male-male competi-tion or aids in grasping females during mating, whereas females may have similar integumental modifications to defend against unwelcome mating attempts by males. Various hairs, pits, bristles, and other integumental fea-tures of arthropods also aid both sexes in sending and detecting acoustic, vibrational, and chemical mating signals (pheromones). The last also oc-curs in loriciferans, where males have specialized spiny scales called *trich-scalids* that are thought to detect pheromones from females.

The functional significance of many other integumental dimorphisms remains a mystery. For example, males and females often differ in the distribution and abundance various surface features such as cilia, hairs, tubercles, papillae, pits, pores, ridges, spines, or scales in body areas that do not have contact during mating. These are usually quite subtle dimor-phisms, at least to our eyes, although they are useful to zoologists as clues to tell the sexes apart. It does seem likely that they serve some biological function, but in most cases we simply know too little about the biology of the organisms to be able to discern what these functions might be.

Sexual Dimorphisms in Color (Sexual Dichromatisms)

Sexual dimorphism in color or pigment patterns (called sexual *dichro-matism*)[26] is familiar to most of us because we see it in many of the do-mesticated and wild species that we frequently encounter. For example, we are used to seeing highly dichromatic birds, especially among the songbirds, gamebirds, and ducks.[27] Many well-known aquarium fishes

such as guppies, Siamese fighting fish, and swordtails also show dramatic
sexual dichromatisms with brightly colored males and dull females. Many
of the common lizards that we find perched on warm rocks are also dichromatic with males eagerly showing their bright throats or underbellies
even in response to human intruders.[28] Among invertebrates, jumping
spiders, damselflies, dragonflies, and butterflies provide the most familiar examples of sexual dichromatisms with males typically more brightly
colored than their mates. These familiar examples may lead us to believe
that sexual dichromatism is very common in the animal kingdom, but
in truth it is the least common of my five categories of secondary sexual
dimorphisms, having been noted in only 26 percent of the animal classes
containing dioecious species (figure 11.1). It occurs with the highest frequency in deuterostomes (41 percent), where it is restricted mainly to
vertebrates, and in the ecdysozoans (27 percent), where it is restricted to
arthropods, especially insects, spiders, and crustaceans. In the Lophotrocozoa sexual dichromatism is uncommon (only 12 percent of classes) and
usually quite subtle, and in the nonbilaterian classes, it is found in only
a few species of sea anemones, box jellies, and sponges, although these
represent 38 percent of classes.

   To understand why sexual dichromatisms are so uncommon and so un-
evenly distributed, we need to understand why animals have colors and
patterns at all. The most universal function of external pigmentation is
to provide protection from ultraviolet radiation, and this is the reason
why the upper or dorsal surfaces of animals tend to be darker than their
lower or ventral surfaces, which face away from the sun. Pigmentation can
also facilitate thermoregulation, particularly in terrestrial species, because
darker colors absorb heat, whereas lighter colors reflect it. Colors and patterns can also reduce the risk of predation by providing camouflage or by
warning predators that the prey animal is noxious, toxic, or venomous.
The bright, conspicuous colors of monarch butterflies, milkweed bugs,
coral snakes, and poison-dart frogs are all examples of the latter adaptation. Other species avoid predation by mimicking the conspicuous warning coloration of these noxious or dangerous species. Milk snakes and king
snakes mimicking coral snakes are a familiar example of this strategy.[29]

Conspicuous coloration can also communicate information between members of the same species. This use of visual signaling[30] is most important for species that engage in group behaviors such as flocking or schooling or in any other forms of social interaction such as territoriality or courtship. In the latter context bright, conspicuous coloration often signals sexual maturity, readiness to mate, and even the health and condition of the signaling individual. Color patterns associated with reproductive function and mating are more likely than other forms of pigmentation to differ between sexes, and as we shall see, sexual signaling is the most common explanation for sexual dichromatism.

The most widespread pattern of sexual dichromatism is for males to be brightly colored and conspicuous while their mates are dull or cryptically colored. This pattern predominates in birds, reptiles, amphibians, fishes, insects, jumping spiders, millipedes, and many crustaceans, and we have abundant evidence that the flamboyant coloration of males in these species can be attributed to sexual selection.[31] Typically, males with more colorful or conspicuous displays have higher mating success than the duller males. This may be simply because they catch females' attention from farther away or stand out from among the competing males. However, females may also be using the color patterns as clues to the health of their potential mates. In many species healthy and well-nourished males tend to have larger, brighter, and more intensely colored displays than their less healthy confreres, and females are able use these displays to select healthy, robust mating partners.[32] Many other sexually dimorphic traits, including body mass and the sizes of ornaments and weapons, also tend to depend on condition, and females use these cues in a similar way.[33] However, color displays seem to be particularly sensitive to the current health and nutritional status of males and so provide particularly effective cues for female choice.

Although sexual selection through female choice has received the most attention as the driving force behind the evolution of sexual dichromatism, males also use color cues in their interactions with each other. For example, male butterflies and dragonflies energetically chase each other while patrolling for females or defending perches, flashing their wing colors as they do so; male red-wing blackbirds flash their red epaulets at each other as they establish their mating territories; male side-blotched lizards energetically display their brightly colored throat patches to establish dominance in their reproductive hierarchy; and male stickleback fishes inspect each others' red throats in contests over nest sites. The relative importance of male-male competition versus female choice in the evolution of sexual dichromatism probably varies greatly among species, but in those listed above, male-male competition seems to be equally or more important than female choice.[34]

There can be little doubt that sexual selection in one form or another explains the bright coloration of males in most sexually dichromatic species, but that is only half the story. We must also ask why the females in these species are dull and inconspicuous. The primary answer is that females reduce their risk of predation by being inconspicuous and blending into the background, that is, by being cryptic. Remaining inconspicuous is particularly advantageous if females brood their young or provide other forms of extended care of offspring because in those species both mothers and their offspring share the predation risk. In species with high levels of male-male aggression, the dull coloration of females may also reduce misdirected male aggression against them. This pattern of sex-specific selection is remarkably consistent across diverse classes and phyla and provides the most general explanation for sexual dichromatisms: males are brightly colored because of sexual selection and females are less brightly colored because of selection favoring inconspicuousness.[35]

Despite this overall trend there are scattered examples of dichromatic species in which females are the more conspicuous sex.[36] We understand this pattern of "reversed" sexual dichromatism less well, but zoologists posit similar underlying causes. In most cases bright coloration in females is thought to be the result of sexual selection on females, and dull coloration in males probably reflects selection for inconspicuousness. We know that males in many species do discriminate among potential mates, preferring females at the peak of their reproductive cycle, in good condition, and with the highest potential fecundity. Males typically use cues such as body size, body shape, and the presence of reproductive pheromones to select the most promising mates, but color is also used as a cue in a number of lizards, birds, fishes, insects, and crabs.[37] In these species, bright female coloration is favored by sexual selection through mate choice. In some species bright colors are also favored by direct female-female competition during territorial disputes or contests for dominance, just as we saw in males.[38] Through either of these mechanisms the more brightly colored females benefit by being able to mate with more males, higher-quality males, or males that provide more resources or better paternal care.[39] The absence of sexual selection favoring bright coloration in males is an obvious requirement for this type of dimorphism, but that is not in itself sufficient. Something has to actively select against bright coloration in males. Perhaps they need to be inconspicuous or cryptic to avoid predation because they do most of the mate searching, as is the case in most of these species. Being cryptic or inconspicuous would also be favored if males provide parental care, especially exclusive parental care.[40] These are plausible explanations, but as yet they have not been rigorously tested. It would be useful to evaluate them across a range of animals that show this pattern of reversed sexual dichromatism, particularly in some of the lesser-known animal groups such as box jellies, ribbon worms, chitons, and marine snails.

Even less common than reversed sexual dichromatisms are color dimorphisms in which the sexes differ in color but in which both are either brightly colored or cryptically colored. Sexually dimorphic camouflage patterns are found in some isopods and shrimp,[41] and sexually dimorphic conspicuous coloration occurs in some in fiddler crabs and *Papilio* butterflies.[42] Each of these is exceptional in a different way. For example, in some fiddler crabs sexual selection plays a major role and favors bright colors in both sexes, whereas in certain *Papilio* butterflies male colors serve primarily for species recognition, and female colors mimic those of co-occurring species that are distasteful or noxious to predators. In the shrimp and isopods both sexes are inconspicuous, but sex-specific habitat preferences or body sizes dictate sex-specific patterns of camouflage. These unusual examples reinforce the impression that we gained from looking at the more typical patterns of dichromatism. It seems that sexual dichromatisms typically reflect differences between the sexes in the relative importance of using color to attract mates or intimidate rivals (sexual selection) versus remaining cryptic to avoid predation and (less commonly) unwanted sexual harassment.

Having explored the reasons why males and females differ in color, you might wonder why sexual dichromatism is restricted to so few animal taxa (only eight phyla and nineteen classes). The main explanation for this is that color is only useful for camouflage or for conveying information to other animals if other animals can perceive it. The intended perceiver must have a visual system capable of detecting the color or pattern.[43] Visual perception varies enormously among animals. In their excellent book on eye design and function, Michael Lande and Dan-Eric Nilsson[44] state that light detectors of any form are completely lacking in about one-third of animal phyla, and in another third, all or almost all species have only simple light detectors that are capable of sensing the amount of light but not its pattern or spectral quality (color). Although most species in the remaining one third of phyla have eyes capable of detecting images and resolving patterns, they vary greatly in their ability to resolve images, detect pattern, or perceive color. The most advanced visual acuity is found in species with large compound eyes or camera eyes that incorporate lenses and mirrors for increased visual acuity. Such eyes have evolved independently in many different animal lineages, including box jellies, cephalopod mollusks, crustaceans, insects, spiders, polychaete worms, and vertebrates. However, although these animals can readily detect patterns, they may not be able to discern true colors (as opposed to shades of gray as in a black-and-white photograph).

Detection of color requires a minimum of two types of light detectors with different spectral sensitivities. Humans, for example, have three types of retinal cone cells for detecting color. These have maximum sensitivities to light with wavelengths of approximately 430, 540, and 570 nm, which correspond to the colors that we see as blue, green, and red, and when combined, they allow us to perceive colors with wavelengths ranging from about 400 to 700 nm. Most mammals other than primates lack the red-sensitive cone type and are effectively red-green color blind.[45]

Plate 1. Northern elephant seals at the Piedras Blancas rookery in California. A large, male harem master (center) is surrounded by his harem of females and their pups. The male is in typical display posture, reared up on his front flippers and vocalizing loudly through his elongated nose. Note the pink wounds on his nose, neck, and chest resulting from physical contests with other males. Photograph by Derek Roff.

Plate 2. Breeding adult female (left) and male (right) southern elephant seals rest during the mating season at the Sea Lion Island rookery in the Falkland Islands. Note the male's large body size and fleshy, elongated nose. Photograph by Tamás Székely.

Plate 3. Female northern elephant seals contest for space in the crowded breeding rookery at Piedras Blancas in California. Photograph by Derek Roff.

Plate 4. A dominant male northern elephant seal attempts to copulate with a female within his harem at the Piedras Blancas rookery in California. The size of her pup suggests that the female may be near the end of lactation and could be fertile. Nevertheless she is protesting vigorously as the male bites her neck, pushes his weight down on her back, and holds her with his left front flipper. Photograph by Derek Roff.

Plate 5. A northern elephant seal harem master (right) at the Piedras Blancas rookery challenges a large rival (left) who has approached too closely to a female on the edge of the harem. The female is croaking loudly to signal her distress at the approach of the males. Note the presence of three smaller males in the intertidal area beyond the edge of the harem. Photograph by Derek Roff.

Plate 6. A great bustard male in breeding plumage. Note his brilliant white neck and under-tail feathers, long moustachial feathers, and the deep chestnut brown at the base of the neck. Photograph by Franz Kovaks.

Plate 7. Great bustard males contesting for dominance on the lek. Photograph by Franz Kovaks.

Plate 8. A great bustard male showing his full breeding display. Photograph by David Tipling.

Plate 9. A great bustard female with her chick. Photograph by David Kjaer.

Plate 10. Adult female (left) and male (right) great bustards. The female is approaching the male aggressively in defense of her nest, and the male is performing a defensive display. Photograph by David Kjaer.

Plate 11. A large, colorful, territorial male *Lamprologus callipterus* displaying to a small, mottled reproductive female. Photograph by Melchior W. N. de Bruin.

Plate 12. A male *Lamprologus callipterus* guarding his shells on the lake bottom. Note the absence of shells in the surrounding areas. A female can be seen inspecting a large shell on the right side of the nest. Photograph by Ad Konings.

Plate 13. A mature female *Argiope aurantia* (left) hanging at the hub of her orb web and accompanied by a mature male (right). Photograph by Troy Bartlett.

Plate 14. Close-up of the front of a mature male *Argiope aurantia* showing the large, paired palps. Note the heavily sclerotized (dark) embolus tips that will break off within the female's reproductive tract. Photograph by Tom Murray.

Plate 15. Female blanket octopus (*Tremoctopus violaceous*) seen from below as she swam at a depth of about 14 m in the Atlantic Ocean near Pompano Beach, Florida. Photograph by Cassandra L. LeMasurier.

Plate 16. Male blanket octopus (*Tremoctopus sp.*) from the Great Barrier Reef, Australia. Photo by D. Paul, from Norman et al. (2002).

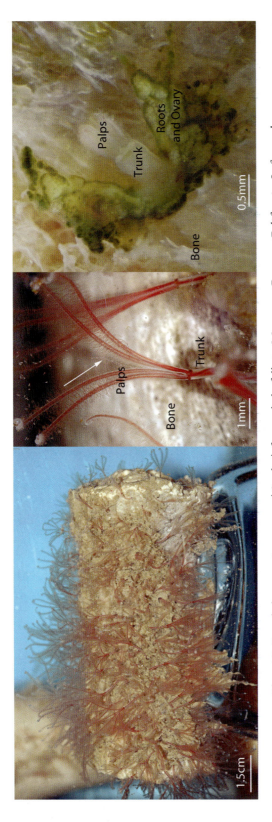

Plate 17. Female bone-eating worms (*Osedax*) from whale-falls in Monterey Canyon, California. Left panel: Numerous *Osedax* "orange-collar" living on (and in) the bone of a gray whale from 600 m depth but being kept in an aquarium. Middle panel: An *Osedax* "orange-collar" female spawning fertilized eggs in the laboratory. Each of the four red palps contains large blood vessels that connect with the main trunk. Note the transparent oviduct (arrow) through which fertilized eggs are transported to be released into the water. Right panel: An *Osedax* "yellow-patch" female showing contracted palps and trunk and with overlying bone dissected away to reveal the ovary and roots. The base of the trunk, the ovary, and roots are sheathed in greenish trophosome tissue. Photographs courtesy of Greg Rouse (Scripps Institution of Oceanography).

Many other animals, including insects, spiders, crustaceans, fish, amphibians, reptiles, and birds are able to detect even shorter wavelengths than mammals and literally see in the ultraviolet range, and some fish and butterflies are able to see longer wavelengths, extending their vision into the near infrared.[46] Not surprisingly the animal groups in which sexual dichromatism is common and striking stand out as having particularly good visual acuity combined with well-developed color vision. For example, among the mammals primates have the most colorful sexual dichromatisms and also the best color vision. Birds, the most colorful and sexually dichromatic of vertebrate classes, have exceptional color vision (five cone types) combined with high visual acuity. Among the spiders wolf spiders are noted for their dramatic sexual dichromatisms and also for their unusually good color vision (they have four types of cone cells, whereas most other spiders have only two). Among insects butterflies, with five cone types, have exceptional color vision that enables them to see a broader spectrum of colors and to discern finer gradations of color than other insects, which typically have only two or three cone types. Dragonflies are also noted for their sexual dichromatism and have four cone types. Perhaps the most compelling example of the juxtaposition of visual perception and sexual dichromatism comes from the cnidarians (jellies). Box jellies are the only cnidarians that have sexual dichromatism, which takes the form of light-colored spots on the female's mantle. They are also the only cnidarians with well-developed camera-type eyes capable of discerning patterns, although they have only one color receptor type and hence are color blind.[47] Unlike other cnidarians, which reproduce by free spawning, male box jellies seek out females and place sperm packets under their bells. It seems likely that the color spot dimorphism in these animals has evolved to enable males to identify females as they swim through the open water and has co-evolved with their distinctive and anomalous complex eye.

The usefulness of color in camouflage or as a signal to other animals is limited not only by the perceptual capabilities of the intended receiver but also by the spectral quality of the environment.[48] Not surprisingly, sexually dichromatic animals tend to be *diurnal* (active in daylight) and to live in brightly lit environments. The aquatic organisms that are sexually dichromatic live in clear, shallow waters or occupy the upper layers of the open ocean (although those in deeper waters often have sexually dimorphic bioluminescence). Terrestrial organisms that are sexually dichromatic live above the surface and often display their colors on brightly lit perches or patches of open ground. We would not expect, nor do we find, marked sexual dichromatism in animals that live their lives in habitats that receive little or no sunlight such as deep or murky water, shaded forest floors, caves or burrows, or soils. Species recognition and sexual signaling in these environments depend on tactile, vibrational, chemical, and acoustic cues rather then visual displays. Given these constraints it is not surprising that sexual dichromatisms are the least common of the types of morphological dimorphisms in my classification. They are probably somewhat more common than we perceive because some may be expressed in the UV or far-red range or as patterns of polarized light that are visible to many animals, but not to us. Nevertheless, they are clearly not as pervasive as dimorphisms in size, shape, appendages, or primary sexual traits.

That concludes my survey of the typical patterns of sexual dimorphism for various categories of sexually dimorphic traits, so it is time to step back and look once again at the overall picture across the animal kingdom. The main conclusion that emerges is that the division of reproductive function between males and females is almost always associated with visible differences in morphology between the sexes. Some form of externally apparent sexual dimorphism has been described for 95 percent of the classes that contain dioecious species,[49] and of these, 84 percent show dimorphism of more than one type, and 65 percent show three or more types of dimorphisms. Dimorphisms in genital openings, body size, and body shape are the most common, but appendage dimorphisms are common in animals with appendages, dichromatism is common in animals with color vision and good visual acuity, and integumental dimorphisms also occur in many of the bilaterian classes. The net result is that males and females in most animal species can be distinguished by an array of morphological traits.

Sexual dimorphisms are not only common, they are also often pronounced. The most extreme differences are found in species with dwarf males, which we found in twenty-three different classes scattered through twelve phyla. In these extraordinary species dramatic differences in body size are almost always accompanied by equally dramatic differences in shape, appendages, and integument, and most dwarf males also show marked reductions in overall morphological complexity. In most animals the differences between the sexes are not as extraordinary as this, but moderate to extreme sexual differences (scores of 3 or 4 in appendix B) are not uncommon. My ranking of moderate refers to species in which the sexes differ so much that they would probably be described separately in field guides, as are male and female birds in many species; so this is a considerable magnitude of sexual dimorphism. Such species occur in eleven different phyla, including the five phyla with the greatest numbers of species: the arthropods, chordates, nematode worms, segmented worms, and mollusks.

Although the types of sexual differences described in the zoological literature are very diverse, in most cases it has been relatively easy to discern the connections between the dimorphic traits and their sex-specific reproductive functions. Males often have genital openings, body shapes, appendages, or integumental features modified for grasping females and delivering sperm, whereas females are likely to have modifications of their body size, shape, and appendages for storing or brooding eggs or offspring (where males participate in care of young, they may have these modifications as well). Appendage, integumental, and color dimorphisms often enable one sex to detect the other at a distance by sending and receiving signals that travel long distances (e.g., sound, pheromones, or visual signals). Body size, shape, and appendage morphologies may all function to increase the efficiency of travel while an animal is searching for a mate. All of these traits are likely to be subject to sexual selection through mate choice or direct competition for mates, and this is a major contributor to the pattern of sexual differences, especially in vertebrates and arthropods. The trait that differs most consistently between the sexes is body size, followed closely by body shape, and in the majority of species, things sort out so that females are larger and thicker-bodied than males, often extraordinarily so. However, if sexual selection strongly favors large males and fecundity selection on females is weak (as in the great bustards, elephant seals, and shell-carrying cichlids), males may be larger and more robust than females, although never more than modestly so when compared to the extremes of female-larger dimorphisms.

CHAPTER 12

Concluding Remarks

The overall patterns of sexual differences summarized in chapters 2 and 11 and the truly extraordinary sexual differences described in the intervening chapters provide rich fodder for thinking about animals and, in particular, for thinking about what it means to be male and female in the animal kingdom. The main impression that I hope you will take away from these chapters is that sexual differences are a major component of the fabric of animal variation. Males and females differ in externally obvious ways in almost all animal classes that contain dioecious species, and in many species the differences are sufficient to be noticed by even a casual observer. Most of us, biologists included, are guilty of typological thinking about animal species. That is, we tend to assign a set of general traits to each animal species (the traits described in field guides and taxonomic keys), and we think of these as relatively invariant among all members of that species. Even where the morphological differences between the sexes are obvious enough to be illustrated in the field guides or noted in the keys, we seldom pay attention to the implications of these differences for aspects of the animal's biology and ecology other than mating interactions. What the preceding chapters have illustrated is that differences between the sexes run deep and often include differences not only in morphology, but also in life history, behavior, and ecology. In fact these more cryptic differences are present even in species where externally apparent morphological differences are slight. The deer mice with which I began this book are good examples of this. How many of us think about sexual differences when we think of mice, and yet for the mice themselves, these differences are huge.

Having convinced you (I hope) that most male and female animals differ fundamentally and often in remarkable ways, let me turn to the rather more elusive issue of how they differ. The clear message from my surveys and examples is that there is no universal pattern of sexual differentiation among animals. Beyond the basic mechanics of producing either sperm or eggs (the single defining feature of being male or female), almost every other aspect of the biology, ecology, life history, and behavior of males and females is up for grabs when it comes to sexual differences. In some species females are giant, lethal predators, and their males are tiny parasitic dwarfs, whereas in other species males are large, belligerent bullies that obtain their mates by force. Sometimes males offer resources or protection for their mates (as in shell-carrying cichlids and elephant seals), but sometimes they seem to offer nothing but their genes (as in great bustards). In some species successful males mate with many different females, and females mate with only one or at most a few males, whereas in other species females accumulate many mates, and males might count themselves lucky to find even one female. Parental care is rare and is almost always the female's role, but in some species both parents care for the eggs or young, whereas in others this role is taken exclusively by males.

The sexes tend to be most similar in species that release their eggs and sperm directly into the environment with no courtship or sexual contact between the spawning individuals. In many such species only gonadal tissue separates the sexes, and at the extreme, in dioecious sponges, the sexes are only distinguishable by gamete type. Sexual differences also tend to be slight in many species where both sexes provide extensive care for the offspring. This is rare, but colonial nesting seabirds such as penguins and gannets are familiar examples. However, the vast majority of species fall between these two extremes and have more obvious sexual differences. Most commonly, parental care is nonexistent or is provided exclusively by the female, and the sexes live apart, coming into close physical proximity only for mating. In these species female morphology typically reflects specialization for egg or offspring production and, less commonly, for parental care. Large body size, thickened bodies, or brood pouches, and cryptic coloration are the most common of these female morphologies. In contrast male morphologies typically reflect adaptations for finding females or for competing with other males to achieve matings and to fertilize eggs. These male morphologies are extraordinarily varied. Various appendages may be modified to serve as weapons, organs of intromission, or organs that send courtship signals. Male bodies may be large and robust for success in physical contests with other males, flamboyantly colored to attract females and intimidate rivals, or small and cryptic if the main task of males is to find rare and elusive females. The enormous variety of male morphologies, life histories, and reproductive tactics is truly staggering.

The net result of all this variance in reproductive tactics and morphologies is an enormous variation in both pattern and magnitude of sexual differences across the animal kingdom. Some portion of this variation is captured by the variance in sexual size dimorphism, which ranges from males that average close to thirteen times heavier than females to females that average hundreds of thousands of times heavier than males. However, as we have seen, these size differences tell only part of the story. Sexual differences are common in virtually every aspect of external morphology, behavior, and life history, and these differences show at least as much variance among species as sexual size dimorphisms. The enduring message from all of this is that there is clearly no one way of being a male or a female animal. Although we can say with confidence that male and female animals are likely to differ in many ways other than primary sexual differences and that these differences almost always reflect specialization for male or female reproductive roles, there is no "normal" or "typical" pattern of sexual differentiation across the animal kingdom.

Our experience as humans tends to bias our perception of what is typical of animals in general. Because we focus very much on our own sexual differences and on the patterns we see in the relatively large, primarily terrestrial animals that we meet in the course of our day-to-day travels, we are likely to think that the common mammal or bird pattern of sexual dimorphism is normal or natural for all animals. I hope that I have succeeded in convincing you that this is a misperception. Our corner of the animal diversity universe is a small one, and it is not at all representative of the vast majority of animal species. Vast numbers of truly extraordinary species escape our notice because they are tiny or live in habitats that we rarely visit and seldom think about. Many of these weird and wonderful creatures have characteristics that we could never have imagined prior to their discovery, including sexual differences that greatly exceed anything found in our usual comfort zone. I have described animals that live on decaying whale carcasses, in the dark ocean depths, in vast reaches of the open ocean, and burrowed inside dead snail shells, but these unusual habitats only touch on the diversity of places that animals live. Soils and lake and ocean sediments are crawling with tiny animals from thousands of species; vast numbers of barely visible animals drift endlessly in the waters of virtually every body of fresh or salt water; and thousands of species from many different phyla live commensally or as parasites on or inside the bodies of other animals. Every terrestrial and aquatic habitat has its share of animal species, most of which are so small that they are at best barely visible to the naked eye. Only a handful of animal classes are familiar to most of us, and almost all of these are either chordates (mammals, birds, reptiles, amphibians, and fishes) or arthropods (insects, spiders, and crustaceans). Diverse as these animals are, they hardly represent the vast array of animal types. We may have a passing familiarity with animals in other phyla such as mollusks (snails, oysters, clams, squid, and octopuses), annelid worms (earthworms and leeches), echinoderms (sea stars, urchins, and sand dollars), cnidarians (jellies, corals, and anemones) and sponges, but it is unlikely that most of us could describe the life histories of any of these animals, let alone imagine how the sexes might differ. The surveys and specific animal stories that populate this book have only touched on the extraordinary diversity of animal life and the many different ways that male and female animals partition their reproductive roles. Nevertheless, I hope that I have been successful in sparking your curiosity about the hundreds of thousands of animal species that are neither similar to us nor part of our everyday experiences and hence normally escape our notice. At the very least we should be cautious about drawing generalities about sexual differences from the relatively few kinds of animals that we know well.

As a final exercise before leaving the topic of sexual differences in the animal kingdom, it seems pertinent to see where we humans fit on the grand scale of animal diversity. How do our sexual differences compare with those we have seen in other animal species? Humans show a fairly typical pattern of sexual dimorphism for a large, terrestrial mammal, with males maturing later and at a slightly larger size than females, and female shape being adapted for gestation, birth, and parental care. Were I to include humans as a separate entry in the table of sexual dimorphisms in appendix B, I would list dimorphisms for visible gonads (male scrotal sacs), genital openings, body size, body shape, and integument (males are more hirsute, especially on the face). Males also have slightly darker hair and skin than females, but this difference is so slight that a zoologist describing our species would probably not notice it, so I would not enter the symbol for color dimorphism. If I were to rank the overall magnitude of our sexual differences using the same scale that I applied to other species, I would give us a score of 3, meaning that our sexual differences are great enough that they would probably be noted in a field guide, but they are not sufficient to be considered extreme. The field guide entry for humans would probably note that males tend to be slightly taller and more robust than females but that the main distinguishing features are the presence or absence of three traits in mature adults: thick facial hair and a penis for males and permanently enlarged mammary glands for females. All three of these traits are very visible signals of reproductive maturity and have doubtless been strongly influenced by sexual selection on both sexes during our evolutionary past, as have body size and shape.[1] We are actually somewhat less sexually dimorphic in body size than would be expected for a primate of our size, which is consistent with both the low degree of polygyny in most human populations and with the presence of at least some degree of biparental care.[2] Our life history, with females maturing earlier and at a smaller size than males, is typical of animals that have only one or two offspring per parturition and rely on repeated reproductive events to augment their lifetime fecundity, especially when this is paired with sexual selection favoring large size or behavioral dominance in males. Overall, our sexual differences in morphology and life history run true to expectations for a mammal of our size, with our mating system, and with extended parental care. They are of sufficient magnitude not to be ignored, but they are certainly not remarkable in comparison to many other animal species, including others in our own clade.

Although our sexual differences are only modest when placed on the scale of animal diversity, we are acutely aware of them in our own lives, and our sex or gender is likely to influence our cultural and social interactions as much or more than it influences our biology. Often our basic biological differences are reinforced and exaggerated by our culture so that even minor, average differences are treated as though they were fixed, dichotomous traits.[3] This does not reflect biological reality. With the exception of our reproductive systems, all humans have the same basic body plan, the same number of parts, the same organs, tissues, and cell types. The nonreproductive traits that do differ between men and women usually differ only in degree so that the sexes differ with respect to the amount or size of the trait, rather than its presence or absence. For example, both sexes have body and facial hair, but it is generally thicker and darker in men. Even when traits differ on average between men and women, the trait distributions of the two sexes typically overlap considerably. The height distributions of American adults from a 1966 census illustrate this well: although the men in this census averaged about 13 cm (5 inches) taller than women, a difference of about 8 percent, only 9 percent of men were taller than the tallest woman and only 5 percent of women were shorter than the shortest man.[4] Most of the nonreproductive physical, physiological, and behavioral characteristics that we consider to be sexually dimorphic in humans show at least this much overlap between the sexes. In comparison there are many animal species in which adult males and females are so different that their body morphologies, life histories, and behaviors are truly dichotomous (i.e., the distribution in one sex does not overlap that of the other), and we have seen a number of such examples in the preceding pages. We are certainly not one of those species. I am a biologist, not a social scientist, and I will leave the debate about the cultural reinforcement of human sexual differences to other authors and other books.[5] My only contribution in this book is to place humans within the grand panorama of animal life. From that perspective our sexual differences are predictable and understandable as biological adaptations to sex-specific reproductive functions, and they are not in any way extreme. There are many more stories waiting to be told about truly extraordinary differences between males and females among the 1.4 million species in the animal kingdom, but the human story is not one of these. In this sense, at least, we are simply not extraordinary.

ACKNOWLEDGMENTS

This book has been a labor of love. It has given me an opportunity to immerse myself in the scientific literature describing the behaviors, morphologies, and life histories of some of the most extraordinary creatures ever found on our planet. A comprehensive book such as this relies on the labor of countless researchers who have observed animals in the wild, conducted experiments, and written hundreds of scientific articles and books documenting their work. I am grateful to all of the authors cited in my sources list, without whom I could never have slaked my curiosity about sexual differences in animals. I am also very grateful to the many talented nature photographers and scientists who have allowed me to use their photographs in the book. My animal portraits come to life through their photographs. I particularly thank Greg Rouse for preparing a series of photographs specifically for my chapter on bone-eating worms and Ted Pietsch for sending me his high-resolution version of the excellent illustration of giant seadevils from Erik Bertelson's monograph. Matthias Foellmer generously allowed me to use some of his unpublished data in my chapter on garden spiders, for which I am very grateful. Of course, I must thank my editor, Alison Kalett, for being endlessly patient and always supportive; without her steady guidance and sage advice, I would certainly have written an encyclopedia rather than a book. I also greatly appreciated the efforts of two anonymous reviewers who carefully read an earlier draft of the book and whose insightful comments helped me to improve the text. Any errors or omissions in the final text are, of course, my own. Finally, I must thank my husband, Derek Roff, who endured far too many evenings and weekends alone while I burrowed down the rabbit hole of my writing. In addition to providing the wonderful photographs that illustrate chapter 3, he has been a sounding board and a source of encouragement throughout the project, and I am deeply grateful, as always, to have him by my side.

Scientific Names Corresponding to Common Names Used in the Text

Common name	Scientific name
Acorn barnacles	Crustacean arthropods in the class Maxillopoda, subclass Cirripedia and order Sessilia
Anglerfishes	Fish in the teleost order Lophiiformes
Arrow worms	Predatory marine worms in the phylum Chaetognatha that are a major component of plankton worldwide
Atlantic cod	*Gadus morhua*
Atlantic salmon	*Salmo salar*
Belostomid water bugs	True bugs in the insect order Hemiptera and family Belostomatidae
Bighorn sheep	*Ovis canadensis*
Birds of paradise	Bird species in the family Paradisaeidae, order Passeriformes
Black grouse	*Tetrao tetrix*
Black widow spider	Several species of spiders in the family Theriidae, genus *Latrodectus*, known as widow spiders. The name black widow applies to at least three North American species (*L. variolus*, *L. mactans*, and *L. hesperus*) and one European species (*L. tredecimguttatus*).
Blanket octopus	*Tremoctopus violaceous*
Bluegill sunfish	*Lepomis macrochirus*
Bony fishes	Approximately 24,000 species of living fish with bony as opposed to cartilaginous skeletons. Classified in the superclass Osteichthyes and including the ray-finned (Actinopterygii) and lobe-finned (Sarcopterygii) fishes.

Common name	Scientific name
Brown songlark	*Cincloramphus cruralis*
Burrowing barnacles	Crustacean arthropods in the class Maxillopoda, subclass Cirripedia, and superorder Acrothoracica
Cassowaries	Large flightless birds (ratites) in the genus *Casuarius*
Chimeras	Marine fishes with cartillaginous skeletons in the Chordate class Holocephali
Cichlid fishes	Bony fishes in the Class Actinopterygii, order Perciformes, and family Cichlidae
Cobweb spiders	Spiders in the family Theridiidae; includes black widow and redback spiders
Common octopus	*Octopus vulgaris*
Conch	True conches are marine snails (gastropod mollusks) in the family Strombidae, but the term is used more generally to apply to any large marine snails that have a high spire and a shell that is pointed at either end.
Coral snakes	Venomous snakes in the orders *Leptomicrurus, Micruroides, Micrurus,* and *Calliophis* within the family Elapidae
Crab spiders	Spiders in the family Thomasidae
Cranes	Birds in the order Gruiformes and family Gruidae
Damselflies and dragonflies	Insects in the order Odonata
Dappled mountain robin	*Arcanator orostruthus*
Deer mouse	*Peromyscus maniculatus*
Emu	*Dromaius novaehollandiae*
Fiddler crab	Crabs in the genus *Uca*
Florida conch	*Strombus alatus*
Fruitfly	*Drosophila melanogaster*
Gamebirds	Birds in the order Galliformes
Garden spiders	Orb-weaving spiders in the family Araneidae, genus *Argiope*
Giant Pacific octopus	*Octopus dolfleini*
Golden silk spiders	Spiders in the family Nephilidae, genus *Nephila*

Common name	Scientific name
Golden orb-weavers	Spiders in the family Nephilidae
Gooseneck barnacles	Crustacean arthropods in the class Maxillopoda, subclass Cirripedia, and order Pedunculata
Great bustard	*Otis tarda*
Green spoon worm	*Bonellia viridis*
Guppy	*Poecilia reticulata*
Hagfishes	Chordates in the class Myxini
Hermit crabs	Crabs in the Arthropod class Malacostraca, order Decapoda, superfamily Paguroidea
Horned beetles	Beetles in the family Scarabidae
Hymenoptera	Insects in the order Hymenoptera (ants, bees, wasps)
Isopods	Small crustaceans in the class Malacostraca, phylum Arthropoda
Jacanas	Birds in the genus *Irediparra*
Jumping spiders	Spiders in the family Salticidae. This is the largest family of spiders with more than 5,000 species.
Kiwi	Ratites in the genus *Apteryx*
Kori bustard	*Ardeotis kori*
Lampshells	Sessile, bivalved marine animals in the phylum Brachiopoda
Leopard seal	*Hydrurga leptonyx*
Little white-shouldered bat	*Amertrida centurion*
Man o' war jellies	Two species of cnidarians in the class Hydrozoa, order Siphonopora and genus *Physalia. Physalia utriculus* is the Pacific man o' war, and *P. physalis* is the Portuguese man o' war.
Manakins	Neotropical birds in the order Passeriformes and family Pipridae
Mantis shrimp	Crustacean arthropods in the class Malacostraca and order Stomatopoda
Mealworm	Larval form of the beetle *Tenebrio molitor*
Midges	A general term referring to several families of small, two-winged insects in the order Diptera and suborder Nematocera

Common name	*Scientific name*
Milk and king snakes	Snakes in the order *Lampropeltis*
Milkweed bug	*Oncopeltus fasciatus*
Monarch butterfly	*Danaus plexippus*
Monkfish	Marine anglerfishes in the order Lophi-iformes, family Lophiidae, genus *Lophius*
Moss animals	Species in the phylum Ectoprocta; also called bryozoans
Muscovy duck	*Carina moschata*
Mussels	Mollusks in the family Mytilidae
New Zealand brown kiwi	*Apteryx australis*
Northern elephant seal	*Mirounga angustirostris*
Oceanospirillales	An order of bacteria in the phylum Proteobacteria and class Gammaproteobacteria. Some strains are specialized endosymbionts of *Osedax* sp.
Orb-weaving or orb-web spiders	Spiders in the order Araneae and infraorder Orbiculariae; includes about one-fourth of known spider species (>11,000 species)
Ostrich	*Struthio camelus*
Peacock	Male peafowl, *Pavo cristatus*
Phalaropes	Shorebirds in the genus *Phalaropus*
Poison-dart frogs	Frogs in the order Anura, family Dendrobatidae
Prairie chicken	*Tympanuchus cupido*
Ratfishes	Marine fishes with cartilaginous skeletons in the Chordate class Holocephali
Ratites	Large, flightless birds in the order Struthioniformes
Ray-finned fishes	Bony fishes in the class Actinopterygii
Redback spider	*Latrodectus hasselti*
Red-wing blackbird	*Agelaius phoeniceus*
Rheas	Ratites in the genus *Rhea*
Rotifers	Members of the phylum Rotifera; also called wheel animals.
Sage grouse	*Centrocercus urophasianus*
Sea anenomes	Benthic, largely sedentary invertebrates in the phylum Cnidaria, class Anthozoa, order Actiniaria

Common name	Scientific name
Sea lilies	Marine, benthic animals in the phylum Echinodermata, class Crinoidea that have arms attached to a stalk. Although they can creep along, they are primarily sessile.
Sea snails	Marine mollusks in the class Gastropoda; includes whelks, conchs and other marine gastropods with external shells
Seahorses and pipefishes	Bony fishes in order Sygnathiformes, family Syngnathidae
Seed shrimp	Small arthropods with a bivalved carapace, in the class Ostracoda
Seven-arm octopus	*Haliphron atlanticus:* an Argonautoid octopus in the family Alloposidae
Sheet web spiders	Spiders in the family Linyphiidae
Shorebirds	Birds in order Charadriiformes
Siamese fighting fish	*Betta splendens*
Side-blotched lizard	*Uta stansburiana*
Southern elephant seal	*Mirounga leonina*
Stalk-eyed fly	*Cyrtodiopsis dalmanni*
Steller sea lion	*Eumetopias jubatus*
Stickleback fish	*Gasterosteus aculeatus,* the three-spined stickleback
Stonefish	Marine, bottom-dwelling estuarian and reef fishes in the family Synanceiidae. Also refers to similar species in the family Scorpionidae (sculpins, scorpionfishes, stonefishes).
Swordtail fish	*Xiphophorus heller*
Termite	Insects in the order Isoptera
Tuberous bushcricket	*Platycleis affinis*
Ungulates	Refers to terrestrial, hoofed mammals in the orders Cetartiodactyla (even-toed undulates, excluding whales in the suborder Cetancodonta) and Perissodactyla (odd-toed ungulates)
Water bear	Members of the phylum Tardigrada
Water flea	Cladocerans: small, freshwater crustaceans in the class Branchiopoda

Common name	Scientific name
Weddell seal	*Leptonychotes weddellii*
Whelk	A general term for large, edible sea snails (gastropod mollusks) primarily in the family Buccinidae
Wolf spiders	Spiders in the family Lycosidae; includes about 2,200 species including tarantulas and bird spiders

Summary of Sexual Dimorphisms by Animal Phylum

This summary presents an alphabetical list of the thirty-one phyla[1] of living animals showing the number of taxonomic classes, the approximate number of living species, the prevalence of sexual reproduction, and the prevalence of dioecy among sexual species. Also shown are the prevalence and magnitude of externally apparent sexual dimorphism in dioecious species. Prevalence refers to the proportion of species in which sexual dimorphism is known to occur and is ranked from 0 to 4 (universal). The magnitude of sexual dimorphism is ranked as follows: 0 (none evident), 1 (evident only to an expert), 2 (differences would be noticed by a novice if males and females were placed side by side), 3 (a notable feature of the species that would be mentioned in field guides), or 4 (sexes so disparate that their major, externally apparent morphological features such as size, shape, and color have nonoverlapping distributions). The rightmost column lists the morphological traits that show externally apparent sexual dimorphism in each phylum with abbreviations as follows: Go, genital opening or gonopore (Goi indicates the presence of a genital intromittent organ); Vg, gonads externally visible; S, body size (f, females usually larger; m, males usually larger; fm, pattern varies among classes; d, dwarf males present); Sh, body shape; App, appendages; Int, integument; and Co, color.[2]

		Number of		Prevalence of		Externally apparent sexual dimorphisms		
Phylum	Common names	Classes	Species	Sexual reproduction[a]	Diecy	Prevalence	Magnitude	Traits
Acanthcephala	Thorny- or spiny-headed parasitic worms	3	1,150	4	4	2	1–4	Goi, Sfd, Sh, Int
Annelida	Segmented worms[b]	2	13,300	3	3	3	1–4	Goi, Sfd, Sh, App, Int
Arthropoda	Insects, spiders, crabs, shrimp, centipedes, millipedes, springtails, barnacles, and relatives	15	1,140,000	3	3	3	1–4	Goi, Sfmd, Sh, App, Int, Co
Brachiopoda	Lamp shells	2	350	4	3	1	1	Sh
Cephalorhyncha	Penis worms, kinorhynchans, and loriciferans	3	290	3	3	2	1–2	Goi, Sf, Sh, Int
Chaetognatha	Arrow worms	0	80	4	0	na	na	na
Chordata	Tunicates, lancelets, and vertebrates[c]	14	57,000	3	3	2	1–4	Goi, Vg, Sfmd, Sh, App, Int, Co
Cnidaria	Jellies, hydroids, corals, anemones, and relatives	5	9,070	4	2	1	1	Vg, Sf, Sh, App, Co
Ctenophora	Comb jellies	2	100	4	1	4	2	Vg
Cycliophora	—	1	2	4	4	4	4	Goi, Sfd, Sh
Echinodermata	Sea stars, sea lilies, urchins, brittle stars, sea cucumbers, and relatives	6	6,500	3	3	1	1–4	Goi, Sfd, Sh, App, Int
Echiura	Spoon worms	1	160	4	4	1	1–4	Sfd, Sh, App
Ectoprocta	Bryozoans or moss animals	2	4,540	4	1	3	1	Go
Entoprocta	Goblet worms	0	150	4	3	1	1	Sfm, Sh
Gastrotricha	Hairy-backs	0	725	2	0	na	na	na
Gnathostomulida	Jaw worms	0	80	4	0	na	na	na

Hemichordata	Acorn worms and pterobranchs	2	90	3	3	1	1	Vg, Go, Sf, App, Co
Mesozoa	—	2	85	4	2	2	3	Sf, Sh, Int
Mollusca	Clams, mussels, oysters, chitons, octopuses, squid, snails, and relatives	7	79,400	4	2	1	1–4	Goi, Vg, Sfmd, Sh, App, Int, Co
Myxozoa	—	0	1,250	4	0	na	na	na
Nemata	Roundworms	2	17,300	3	3	3	1–4	Goi, Sfd, Sh, Int
Nematomorpha	Horse-hair worms	0	305	4	4	3	2	Sfm, Sh
Nemertea	Ribbon or proboscis worms	2	1,250	3	3	1	1	Vg, Sfd, Sh, Co
Onychophora	Velvet worms	0	100	3	4	2	1–2	Go, Sf, Sh, App, Int
Phoronida	Horseshoe worms	0	12	3	1	1	1	Sh, App
Placozoa	—	0	1	4	0	na	na	na
Platyhelminthes	Flatworms[d]	3	24,000	3	1	3	1–4	Goi, Sfd, Sh
Porifera	Sponges	3	5,650	4	1	1	1	Co
Rotifera	Rotifers	2	1,853	2	4	3	1–4	Goi, Sfmd, Sh, App, Int
Sipuncula	Peanut worms	0	320	3	3	1	1	Vg
Tardigrada	Water bears	2	780	3	3	2	1	Goi, Sf, Sh, App

[a]Sexual and asexual reproduction (e.g., reproduction by fission, budding, or parthenogenesis) are not mutually exclusive. Many phyla include both asexual and sexual species and also species that use both modes of reproduction. Asexual reproduction is common in eleven of the thirty-one phyla (Cnidaria, Cyclophora, Ectoprocta, Gastrotricha, Mesozoa, Myxozoa, Placozoa, Platyhelminthes, Porifera, Rotifera, and Tardigrada), but rare in thirteen, and absent altogether from seven phyla (Acanthocephala, Brachiopoda, Cephaloryncha, Chaetognatha, Echiura, Gnathostomulida, and Nematomorpha). Even phyla with a rank of 4 for prevalence of sexual reproduction may include species that reproduce asexually as well as sexually.

[b]Segmented worms include earthworms, leeches, tube worms, and polychaete worms.

[c]Vertebrates include fishes, amphibians, reptiles, birds, and mammals.

[d]Flatworms include tapeworms, flukes, and turbellarians.

(continued)

SOURCES: Primary sources: Breder and Rosen (1966), Geise and Pearse (1974, 1975a, 1975b, 1977, 1979), Ghiselin (1974), Blackwelder and Shepherd (1981), Bell (1982), Charnov (1982), Adiyodi and Adiyodi (1989, 1990, 1992, 1993, 1994), Lombardi (1998), Conn (2000), Pechenik (2005), Jarne and Auld (2006), de Meeûs et al. (2007), Hickman et al. (2007), and Bisby (2008). Taxonomy and nomenclature follow the ITIS Catalogue of Life: 2008 Checklist (Bisby, 2008).
Additional sources for specific groups:
Acanthocephala: Poulin and Morand (2000).
Arthropoda: Sharma and Metz (1976), Gilbert (1983), Poulin (1996), Hopkin (1997), Minnelli et al. (2000), Ohtsuka and Huys (2001).
Brachiopoda: James et al. (1991).
Cephalorhyncha: Neuhaus and Higgins (2002), Kristensen (2002).
Chordata: Shine (1979), Breder and Rosen (1966), Sims (2005), Filiz and Taskavak (2006), Cox et al. (2007), Kupfer (2007), Lindenfors et al. (2007).
Cnidaria: Lewis and Long (2005).
Ctenophora: Harbison and Miller (1986).
Echinodermata: Vail (1987), Hamel and Himmelman (1992), O'Loughlin (2001), Stöhr (2001), Emlet (2002), Tominaga et al. (2004).
Ectoprocta: Ostrovsky and Porter (2011).
Hemichordata: Hadfield (1975), Sastry (1979), Heller (1993).
Mollusca: Crozier (1920), Coe (1944), Webber (1977), McFadien-Carter (1979), Pearse (1979), Heller (1993), Gowlett-Holmes (2001), Lamprell and Healy (2001), Lu (2001), Schwabe (2008).
Myxozoa: Kent et al. (2001).
Nemata: Maggenti (1981), Poulin (1997).
Nematopmorpha: Schmidt-Rhaesa (2002), Cochran et al. (2004).
Nemertea: McDermott and Gibson (1993), Roe (1993), Stricker et al. (2000), Döhren and Bartolomaeus (2006).
Phoronida: Temereva and Malakhov (2001).
Placozoa: Pearse and Voigt (2007).
Platyhelminthes: Campbell (1970), Bell (1982).
Rotifera: Gilbert and Williamson (1983), Ricci et al. (1993).
Tardigrada: Claxton (1996), Guldberg and Kristensen (2006), Garey et al. (2008).

NOTES

Chapter 1: Introduction

1. Readers interested in the results of this study of deer mice will find them in Fairbairn (1977a,b and 1978a,b).

2. The most general definition of natural selection is change across generations in heritable biological characteristics caused by consistent relationships between trait values and fitness.

3. Darwin defined sexual selection in *The Origin of Species* (1859) as selection that "depends not on a struggle for existence, but on a struggle between males for possession of the females" (Darwin, 1859, reprinted 1986, pp. 136–38). He greatly expanded on the concept and its implications for humans in his two-volume treatise, *The Descent of Man and Selection in Relation to Sex* (1871).

4. Secondary sexual traits are traits that distinguish males and females within a species but are not components of the reproductive tract (i.e., the gonads, reproductive ducts, and genitalia). Components of the reproductive tract are collectively called primary sexual traits.

5. Clutton-Brock (2009).

6. For examples see Ghiselin (1974), Lande (1980), Arnold (1985), Bradbury and Andersson (1987), Hedrick and Temeles (1989), Berglund et al. (1993), Anderson (1994), Arnold and Duvall (1994), Eberhard (1996, 2009), Fairbairn (1997), Birkhead and Moller (1998), Amundsen (2000), Dunn et al. (2001), Wedell et al. (2002), Badyaev and Hill (2003), Blanckenhorn (2005), Fairbairn et al. (2007), and Cox and Calsbeek (2009).

7. Phyla (singular phylum) are the major evolutionary lineages of animals and the highest level in the formal taxonomic classification of animals.

8. Sexual dimorphism refers to any consistent difference between males and females within a given species. See the glossary for a more extensive definition.

9. Peters (1983), Calder (1984), Schmidt-Nielsen (1984), Reiss (1989), Roff (1992), Stearns (1992), and Brown and West (2000).

10. For discussions and examples of the relationships between sexual size dimorphisms and sexual differences in other traits, see Ghiselin (1974), Ralls (1977), Fairbairn (1997), Vollrath (1998), Weckerly (1998), Blanckenhorn (2005), Fairbairn et al. (2007), and the many more examples in later chapters.

Chapter 2: The Roots of Sexual Differences

1. To determine how common dioecy is across the animal kingdom, I gathered information on the reproductive strategies recorded for species in each animal phylum and its included taxonomic classes (a total of thirty-one phyla and eighty-one classes). Recent advances in evolutionary developmental biology combined with the increasingly refined techniques of molecular systematics have generated major revisions to traditional phylogenetic hypotheses over the past decade (e.g., compare Adoutte et al., 2000; Nielsen, 2001; Tree of Life 2002a,b,c; Philippe et al., 2005). Much controversy still exists, and it is difficult to find two sources that agree on all aspects of the evolutionary tree of animals. I used the thirty-one animal phyla given in the *ITIS Catalogue of Life: 2008 Annual Checklist* (Bisby et al., 2008). Additional sources and a summary of my survey results for each phylum can be found in appendix B, and I have archived more detailed information for each phylum and class in the Dryad repository at http://dx.doi.org/10.5061/dryad.n48cm (Fairbairn, 2013).

2. My discussion of mechanisms of sex determination is based on data from Bull (1980, 2008), Austin and Edwards (1981), Gilbert and Williamson (1983), Petraits (1985), Adiyodi and Adiyodi (1993), Hayes (1998), Lombardi (1998), Marin and Baker (1998), Scherer and Schmid (2001), Simonini et al. (2003), Carrier et al. (2004), Eggert (2004), Oliver and Parisi (2004), Haag and Doty (2005), Ming and Moore (2007), Uller et al. (2007), and Vandeputte et al. (2007).

3. For more information about the development of sexual differences in humans and other animals see Lyon (1994), Marshall Graves (1994), Mealey (2000), and Bainbridge (2003).

4. Acanthocephalan worms, polychaete worms, and roundworms.

5. ZZ/ZO systems are relatively rare but have been found in a few birds and reptiles.

6. This is called haplodiploidy because males have half the chromosome complement of females (ploidy refers to the number of sets of chromosomes).

7. Some polychaete worms, fishes, bivalve mollusks, and possibly sponges.

8. Sex determination by environmental or social cues has evolved independently in many groups, including various species of fishes, amphibians and reptiles, crustaceans, polychaete worms, gastropods (snails and slugs), and possibly bivalve mollusks. Often temperature is the cue that sets the developmental pathway, but other cues such as density, sex ratio, and even body size at a critical

age are used by some species. The example of the green spoon worm comes from Jaccarini et al. (1983) and Berec et al. (2005).

9. For sex-specific patterns of gene expression in humans, fruit flies, and chickens, see Saifi and Chandra (1999), Parisi et al. (2003), Kaiser and Ellegren (2006), and Talebizadeh et al. (2006). For a review and discussion of the role of sex linkage versus sex-specific gene expression in the evolution of sexual dimorphisms, see Fairbairn and Roff (1996) and Mank (2009). For an excellent demonstration of the conservation of sex-specific genetic cascades in spite of wide variation in sex-determination systems, see Ferguson-Smith (2007).

10. For more information on epigenetic effects, see Russo et al. (1996).

11. Degnan and Degnan (2006).

12. See Morrow (2004), Reunov (2005), Quirk (2006), and Pitnick et al. (2009) for sperm counts and discussions of the advantages of large numbers of sperm.

13. The only exceptions are the Brachiopoda, Ctenophora, Cycliophora, Ectoprocta, Entoprocta, Hemichordata, Phoronida, Placozoa, Porifera, and Sipuncula.

14. This pattern is particularly well documented in mammals, birds, fishes, and insects (Stockley et al., 1997; Birkhead, 1998; Simmons, 2001; Pizzari and Parker, 2009).

15. For discussions and more examples of ways in which males compete for fertilization success during copulation and within the female's reproductive tract, see Eberhard and Cordero (1995), Eberhard (1996, 2009), Choe and Crespi (1997), Stockley et al. (1997), Simmons (2001), and Pizzari and Parker (2009).

16. See more examples and discussions of male mating competition in Ghiselin (1974), Parker (1978), Andersson (1994), Fairbairn et al. (2007), and Emlen (2008).

17. Lombard (1998).

18. Ghiselin (1974) and Vollrath (1998).

19. Quirk (2006).

20. My description of chickens and eggs comes from Wilson (1991) and Laughlin (2005).

21. Stearns (1976, 1992), Roff (1992), and Messina and Fox (2001).

22. Calder (1979).

23. For information on cod spawning, eggs, and larvae see Thorsen et al. (1996), Nissling et al. (1998), Hansen et al. (2001), Ouellet et al. (2001), and Fudge and Rose (2008).

24. Estimates of proportional clutch mass come from Roff (1992).

25. Vahed et al. (2011).

26. Roff (1992).

27. Woodroffe and Vincent (1994).

28. Lombardi (1998).

29. For examples of females that brood or guard their eggs, see Blüm (1985), Clutton-Brock (1991), and Lombardi (1998).

30. This social system is called *eusociality*. More complex relationships develop in some species, including the presence of multiple queens, but there is always a distinct caste of sterile female workers that tend the eggs, larvae, and pupae.

31. Males are the larger sex on average in only six animal classes: Mammalia, Aves, Malacostraca (e.g., crabs, lobsters, shrimp), Ostracoda (seed or mussel shrimp), Entoprocta (goblet worms), and Polyplacophora (chitons). Species in which males are the larger sex also occur in some species within eleven other classes where this is not the dominant pattern.

Chapter 3: Elephant Seals

1. This information about the rookery at Piedras Blancas comes from the website maintained by the Friends of the Elephant Seal (FES at http://www.elephantseal.org), a nonprofit organization formed to protect and monitor the colony. The site now has ample parking, an official viewing platform with information boards, and viewing trails. Docents provided by the FES are usually present on site to provide information and assistance.

2. http://www.iucnredlist.org/apps/redlist/details/13581/0a.

3. The information on the numbers and distribution of southern elephant seals comes from McConnell (1992) and the Seal Conservation Society web page at http://www.pinnipeds.org.htm accessed in August 2012.

4. Size ratios of male and female elephant seals differ among colonies. Estimates also depend on which males are included: only males in harems, all breeding males, or all sexually mature males. Because breeding males in harems tend to be the largest males, the estimates of sexual size dimorphism based on these animals are largest. Size ratios also change as the mating season progresses, and both sexes lose weight. Le Boeuf and Laws (1994a) give a range of 1.5–10 for the weight ratios of individual mating pairs of southern elephant seals, but estimates of the ratio of average weights of males and females typically range between 7 and 8, as indicated in table 3.1.

5. See Hindell et al. (1991) and LeBoeuf and Laws (1994a) for explicit comparisons of quantitative ecological, behavioral, and reproductive parameters for the two species. Other good sources of detailed information on elephant seal biology include the edited volume by Le Boeuf and Laws (1994b) and an earlier study of reproductive success in male and female northern elephant seals by Le Boeuf and Reiter (1988). Several websites provide more general overviews as well as up-to-date information about specific populations or rookeries. Information on northern elephant seals can be found at http://www.iucnredlist.org/apps/ redlist/details/13581/0, http://www.pinnipeds.org/species/nelephnt.htm, and http://www.elephantseal.org. Southern elephant seal populations are described at http://www.iucnredlist.org/apps/redlist/details/13583/0 and http://www .pinnipeds.org/species/selephnt.htm. In addition to these sources I have obtained details of elephant seal biology from many individual research papers that are noted throughout the text as well as in table 3.1.

6. My description of female reproductive strategies and behavior on the rookery comes from Le Boeuf and Reiter (1988), Deutsch et al. (1994), Fedak et al. (1994), Le Boeuf and Laws (1994a), Sydeman and Nur (1994), Arnblom et al. (1997), Crocker et al. (2001), MacDonald and Crocker (2006), and Fabiani et al. (2006). Information on survival of mothers and pups derives from those sources as well as Le Boeuf et al. (1994) and McMahon et al. (2000).

7. More than 70 percent of females give birth within 4 km (1.8 mi) of their own birthplace and more than 60 percent are within a few hundred meters (Le Boeuf and Reiter, 1988; Fabiani et al., 2006).

8. Twinning occurs in less than 0.3 percent of parturitions (Arnblom et al., 1997).

9. At Año Nuevo rookery in California MacDonald and Crocker (2006) found that the ratio of pup mass gained to maternal mass lost was only 45 percent for first-time mothers, compared with 61 percent for experienced mothers.

10. At Año Nuevo rookery in California weaning success increases asymptotically with a mother's age, most rapidly between the ages of three and six. In most colonies, there is little or no effect of age beyond eight years (Le Boeuf and Laws, 1994a).

11. In a study of southern elephant seals at Marion Island, Pistorius et al. (2008) found that only 83 percent of first-time mothers survived the two-month postbreeding period of foraging at sea, compared with 91 percent of experienced mothers.

12. Hindell et al. (1991), McConnell et al. (1992), Antonelis et al. (1994), Le Boeuf (1994), Slip et al. (1994), Stewart and De Long (1994), Le Boeuf et al. (2000), and Lewis et al. (2006).

13. In a study of southern elephant seals Galimberti et al. (2000a,b,c) observed that about three-quarters of male mating attempts occurred outside of the female's period of estrus. On average females faced two to three mating attempts per day and were unreceptive almost 87–88 percent of the time. Males sometimes forced copulations on pregnant or anestrus females, but this was uncommon.

14. My description of males in the breeding rookery comes from Clinton (1994), Deutsch et al. (1990), Haley et al. (1994), Le Boeuf and Laws (1994a), Modig (1996), Galimberti et al. (2002a,b, 2007), Fabiani et al. (2004), Carlini et al. (2006), and Sanvito et al. (2008).

15. Also see Haley et al. (1994).

16. Deutsch et al. (1990), Modig (1996), and Galimberti et al. (2000c, 2002a).

17. Fabiani et al. (2004).

18. Sanvinito et al. (2007a,b, 2008).

19. Haley et al. (1994) provide an excellent analysis of the relative importance of size and age in predicting the dominance status of male northern elephant seals, and Carlini et al. (2006) provide similar analyses for southern elephant seals. See Reiss (1989) for a general treatment of the benefits of large size in animals.

20. Clinton (1994).

21. The differences before age two are too small to be distinguished on the scale of figure 3.3, but see table 3.1. Also see growth trajectories for southern elephant seals in Field et al. (2007).

22. Boyd et al. (1994).

23. Le Boeuf and Mesnick (1991), Mesnick and Le Boeuf (1991).

24. Galimberti et al. (2000a,b,c).

25. Female northern elephant seals tend to protest more vigorously and loudly when approached by males other than their harem master, suggesting that they prefer to mate with the latter. This reinforces the sexual selection favoring harem masters derived from direct male-male competition. There is no evidence that southern females discriminate in this way.

26. Deutsch et al. (1994), Galimberti et al. (2000c, 2007), McDonald and Crocker (2006).

27. Galimberti et al. (2007).

28. Alexander et al. (1979), Bonner (1994), Lindenfors et al. (2002).

29. Boyd et al. (1994) measured biomass and energy consumption of male and female elephant seals in the South Georgia population and found that males accounted for 63 percent of the biomass and 59 percent of the annual energy expenditure of the population.

30. The divergence in foraging areas and prey types between sexes may have evolved to reduce competition between the sexes for food and certainly achieves this function in present populations. However, preferences for different food types may also be a secondary consequence of sexual differences in nutritional needs. Similarly, the use of different migration routes, foraging areas, and foraging behaviors may reflect biomechanical and physiological constraints imposed by the different body sizes and growth rates of the two sexes. These possible explanations are not mutually exclusive, and all three processes may be at work. Analogous sexual segregation of habitat use outside of the breeding season occurs in many vertebrate species, across all vertebrate classes. The edited volume by Ruckstuhl and Neuhaus (2005) describes many examples and excellent discussions of the proximal and evolutionary explanations for this common pattern.

31. Bonner (1994).

32. For ungulates see Geist (1971), Hogg (1987), Clutton-Brock et al. (1988), Festa-Bianchet (1991), Hogg and Forbes (1997), Kruuk et al. (1999), Post et al. (1999), Ruckstuhl and Neuhaus (2005), and Pelletier and Festa-Bianchet (2006). For primates see Lindenfors and Tullberg (1998) and Lindenfors (2002).

33. Females usually reinforce this selection for large size by preferring to mate with the dominant male and protesting mating attempts by outside males. Female elephant seals do this mainly by preferring to give birth and nurse their pups in the central part of a harem and by preferring larger harems. Northern elephant seal females sometimes display an even stronger preference by protesting more vigorously when approached by males other than the harem master.

Chapter 4: Great Bustards

1. In the mid-eighteenth century great bustards could be found from Britain in the northwest, south to the Iberian Peninsula and Morocco, and eastward as far as Mongolia and northern China, and they were so common in some parts

of Europe that they gathered in flocks of thousands and were regarded as agricultural pests. However, they have since disappeared from much of their former range. The total world population is estimated to be about 45,000 birds, and the species is not considered to be at risk of extinction, but most of the small populations remain vulnerable. In Britain the last bustard egg hatched in 1832, and the birds disappeared entirely in the 1840s. They have been reintroduced to the Salisbury Plain in England, and the first chicks hatched in the wild there in 2009. This reintroduction is being managed and funded by the Great Bustard Group, a registered international charitable organization whose aim is to protect the interests of the great bustard throughout its range. Information on the GBG and their conservation efforts in the UK and throughout Eurasia can be found at http://greatbustard.org/. For more information on the history and present status of great bustard populations see Johnsgard (1991, 1999), Martín et al. (2001a), Nagy (2007), Alonso, Martín et al. (2009), and Birdlife International (2009) Species fact sheet: *Otis tarda*. http://birdlife.org 20/2/2010.

2. The combination of crane-like and grouse-like traits has made bustards (the family Otididae) difficult to classify, but recent genetic analyses indicate that they are most closely related to cranes (Pitra et al., 2002; Mindell et al., 2008). This description of their diet comes from Johnsgard (1991), Alonso et al. (1998), and Morgado and Moreira (2000).

3. Johnsgard (1991, 1999), and Alonso, Magaña et al. (2009).

4. Székely et al. (2007) give a body mass ratio (males/females) of 3.14. Johnsgard (1991), Hidalgo and Carranza (1991), and Carranza and Hidalgo (1993) say males usually weigh more than three times as much as females. However, Alonso, Magaña et al. (2009) found a mean mass ratio of only 2.48 in a contemporary Spanish population and suggest that very large males are now rare in these populations.

5. Dunn et al. (2001) and Székely et al. (2007).

6. Raihani et al. (2006) and Székely et al. (2007).

7. My description of great bustard mating behavior, including lek formation, male displays, male-male interactions, and female mate choice is compiled from the following sources: Ena et al. (1987), Johnsgard (1991), Hidalgo de Trucios and Carranza (1991), Carranza and Hidalgo de Trucios (1993), Alonso et al. (1998), Alonso, Magaña et al. (2009), Morgado and Moreira (2000), Morales et al. (2001, 2003), and Raihani et al. (2006). Where information is specific to one source, I have indicated this by a separate footnote. Where seasonal timings of events are given, these refer to Spanish populations.

8. Carranza and Hidalgo de Trucios (1993), and Alonso, Magaña et al. (2009). Here condition means weight for a given skeletal size. In natural populations condition is presumed to correlate with good health and with reserves of energy in the form of stored carbohydrates, fats, and protein.

9. Morales et al. (2003).

10. Morales et al. (2003) and Alonso, Magaña et al. (2009).

11. Caranza and Hidalgo de Trucios (1993), Raihani et al. (2006), and Alonso, Magaña et al. (2009).

12. Carranza and Hidalgo de Trucios (1993), Alonso et al. (2004), and Martin et al. (2007).

13. My description of female mating behavior is taken from Ena et al. (1987) and Hidalgo de Trucios and Carranza (1991).

14. Most females nest within about 5 km (3 miles) of the lekking areas, but nests have been found up to 18 km (11 miles) from the closest lek. My description of female nest site selection refers primarily to populations in central and northwestern Spain (Morales et al., 2002; Magaña et al., 2010) and Portugal (Morgado and Moreira, 2000).

15. My description of female egg laying and parental care comes primarily from Ena et al. (1987), Johnsgard (1991), Morgado and Moreira (2000), Morales et al. (2002), Martín et al. (2007), Nagy (2007), and Magaña et al. (2010).

16. According to Johnsgard (1991), young birds eat primarily insects in summer (96 percent of diet is animal origin). Plant material, especially seeds, becomes more important in autumn and dominates in winter. The adult diet is about 87 percent animal origin in summer and 40 percent overall. Most animal food is insects, but birds and small rodents are also eaten.

17. If the female does not nest the following spring, supplementary feedings may continue for as long as seventeen months (Alonso et al., 1998).

18. Alonso et al. (2000).

19. It is likely that the food requirements of the rapidly growing chicks and the resulting dispersion of mothers and chicks across the landscape set the upper limit for population densities (Morales et al., 2002).

20. Johnsgard (1991) and Nagy (2007).

21. Ena et al. (1987).

22. Chicks were considered to have fledged (become independent) at the end of August. More recent studies covering multiple years have found similar or even

lower chick survival rates and female reproductive success in the same region of Spain (Alonso et al., 2000, 2004; Morales et al., 2002; Martín et al., 2007).

23. Morales et al. (2002).

24. Alonso, Magaña et al. (2009).

25. The energy required for animals to fuel their basic metabolic needs increases with body size, and so, as a general rule, larger individuals have to obtain more calories per day to sustain themselves (Calder, 1984; Brown and West, 2000). The positive scaling of food requirements with body size could be offset if foraging efficiency increases with body size so that larger females are able to obtain more food. One could imagine such a scenario for animals that capture large prey that must be chased or physically subdued. Alternatively, larger animals may be able to exclude smaller ones from food resources through territoriality or dominance interactions, allowing the former to obtain a greater share of the food available. Neither scenario would seem to apply to female great bustards, at least during the period when they are foraging alone with their chicks. The trade-off between providing energy for offspring versus self-maintenance is not unique to bustards. On the contrary it occurs in various guises in all female animals and is one of the major determinants of the optimal female body size.

26. Alonso et al. (2004), Martín et al. (2007), Alonso, Magaña et al. (2009), and Alonso, Palacín et al. (2009). For example, annual mortality rates for adults are 8 percent per year for females compared to 13 percent for males.

27. Carranza and Hidalgo de Trucios (1993).

28. In addition to the energetic trade-off discussed previously, recall that fecundity is not positively correlated with body size in female birds (see chapters 2 and 11 for discussions of this).

29. Ena et al. (1987), Alonso and Alonso (1992), Alonso et al. (1998), and Morales et al. (2002).

30. Morales et al. (2002) and Martín et al. (2007).

31. Martín et al. (2007) and Alonso, Magaña et al. (2009).

32. Alonso et al. (1998) also found that chicks of both sexes grew larger if reared on high-quality home ranges and by mothers that were effective foragers. However, male chicks gained an incremental benefit because females that had high foraging success increased their supplemental feeding of male chicks proportionally but did not do so for female chicks.

33. Many studies have reported higher mortality for males than for females. Examples include Martín et al. (2007), Alonso et al. (2004), Alonso, Magaña et

al. (2009), and Alonso, Palacín et al. (2009). Males are particularly susceptible to starvation as chicks and to death by collision with power lines as adults.

34. This description of juvenile dispersal comes primarily from Alonso and Alonso (1992), Alonso et al. (1998), Morales et al. (2000), and Martín et al. (2001a).

35. This description of adult movements comes from Martinez (1988), Johnsgard (1991), Alonso et al. (2000, 2001), Morales et al. (2000), Martín et al. (2001b), Alonso, Palacín et al. (2009), and Palacín et al. (2009).

36. These arguments are a bit tricky because they depend on how thermoregulation, basal metabolic rate, and the amount of energy stored as carbohydrate and fat scale with body mass. Alonso, Palacín et al. (2009) and Palacín et al. (2009) detail the hypothesis for preferential summer migration of males in Spanish populations, and Streich et al. (2006) outline the hypothesis for preferential winter migration of female great bustards from German populations.

37. Höglund and Alatalo (1995) provide an excellent and comprehensive review of lekking in animals that includes a table listing species in which lekking has been well documented. This table includes ninety-seven bird species, thirteen mammal species, eleven amphibian species, twenty-four fish species, and seventy-two insect species. Lekking has also been observed in reptiles, notably marine iguanas (Partecke et al., 2002). Shelly and Whittier (1997) review lekking in insects and add expand the list from Höglund and Alatalo (1995) to include 130 species from eight different orders. See Höglund and Sillén-Tullberg (1994) and Andersson (1994) for excellent reviews of lekking in birds and Clutton-Brock et al. (1993) for a review and discussion of lekking in ungulates. Keenleyside (1991) and Barlow (2000) describe lekking in African cichlids.

38. Clutton-Brock et al. (1993) and MacKenzie et al. (1995).

39. Great bustards are generally silent during their displays, but they are unusual in this.

Chapter 5: Shell-Carrying Cichlids

1. Keenleyside (1991) and Barlow (2000).

2. Standard lengths (length from the tip of the snout to the end of the vertebrae) and sexual size ratios for ninety-nine African cichlid species are given in Erlandsson and Ribbink (1997).

3. Schultz and Taborsky (2000) and Ota et al. (2010).

4. Shell brooding is found in at least fifteen species in Lake Tanganyika and has evolved independently at least four times, including in *L. callipterus* (Sato, 1994; Sturmbauer et al., 1994; Barlow, 2000; Schultz and Taborsky, 2000).

5. Sato (1994), Schultz and Taborsky (2000, 2005), and Ota et al. (2010).

6. Sato (1994) reports a mean of 131 eggs and a range of 39–280 eggs. Survival from egg to independence was 71 percent. Meidl (1999) counted fry ten to fourteen days after spawning and found brood sizes of 4 to 180, with a mean of 97.2.

7. This discussion of female preferences for shell size comes from Schultz and Taborsky (2000, 2005), Taborsky (2001), and Ota et al. (2010).

8. Schultz et al. (2006).

9. Sato (1994).

10. Schultz et al. (2006) and Otah et al. (2010) also demonstrate that fecundity is proportional to the cube of body length, so it increases much more rapidly than length itself.

11. The data are from Sato (1994).

12. My description of the effect of shell size on female size is taken from Sato (1994), Schultz and Taborsky (2005), Schultz et al. (2006), and Ota et al. (2010).

13. My descriptions of the life histories of males and females are taken primarily from Taborsky (2001), Sato et al. (2004), Schultz et al. (2006, 2010), Maan and Taborsky (2008), and Ota et al. (2010). They refer to the Zambian populations at the south end of Lake Tanganyika.

14. My description of male territoriality and shell transport behavior comes from Sato (1994), Schultz and Taborsky (2000, 2005), Barlow (2000), Taborsky (2001), and Maan and Taborsky (2008).

15. My description of courtship and spawning is taken from Sato (1994), Meidle (1999), Barlow (2000), Schultz and Taborsky (2000), and Schultz et al. (2010).

16. My description of male reproductive behaviors is taken primarily from Sato (1994), Barlow (2000), Meidle (1999), Schultz and Taborsky (2005), Maan and Taborsky (2008), Ota et al. (2010), and Schultz et al. (2010).

17. Maan and Taborsky (2008) found that usurping males sometimes fail to identify broods in the early stages of development, and so these may survive a takeover, but broods in later stages (wrigglers) are almost always detected and killed.

18. For analyses of the energetic costs of male mating behaviors see Schultz et al. (2010). Larger males actually lose slightly more mass per day in proportion to their body volume than smaller males, but they can endure fasting longer because they have higher initial condition.

19. These values for shell and male sizes come from Sato (1994) and Ota et al. (2010). The threshold size for lifting shells was determined by Schultz and Taborsky (2005).

20. Ota et al. (2010).

21. Sato (1994), Meidle (1999), Taborsky (2001), and Schultz and Taborsky (2005).

22. See Ota et al. (2010) for an excellent comparison of the effect of shell distributions on male and female body size in seven populations of *L. callipterus*. Sato (1994), Schultz and Taborsky (2005), and Maan and Taborsky (2008) provide detailed descriptions of the individual study populations.

23. The dwarf male tactic is not unique to *L. callipterus*. It is quite common in substrate-breeding fishes, and the males adopting this tactic are often called parasitic males. However, I prefer to reserve the adjective "parasitic" for males that become parasites on their mates, as in the seadevils of chapter 8, and so I have avoided it here.

24. Dwarf males do grow after maturity and average only 3.4 cm long and less than a gram in weight, about 2.5 percent of the weight of territorial males.

25. My description of alternative reproductive tactics in *L. callipterus* comes from Sato (1994), Meidl (1999), Taborsky (2001), Sato et al. (2004), and Schultz et al. (2010). See Taborsky (2001) and Mank and Avise (2006) for excellent reviews of variation in reproductive behaviors in fishes in general, and see Gonzalez-Voyer et al. (2008) for comparisons of cichlid species. Fairbairn et al. (2007) discuss variation in mammals, birds, reptiles, amphibians, spiders, and insects.

26. See Gonzalez-Voyer et al. (2008) for an excellent analysis of the balance between parental care and sexual selection in cichlid fishes that shows how it influences patterns of sexual dimorphism across the Cichlidae.

27. See chapters in Fairbairn et al. (2007) for examples and discussions of the relationship between sexual size dimorphism and sexual selection in mammals, birds, reptiles, and insects. Erlandsson and Ribbink (1997) and Gonzalez-Voyer et al. (2008) consider this for cichlid fishes, and Parker (1992) and Roff (1992) discuss this for teleosts (ray-finned fishes) in general.

Chapter 6: Yellow Garden Spiders

1. The genus *Nephila* is characterized by very large females and extreme sexual size dimorphism. *Nephila turneri* is the most size dimorphic, with females

averaging 32.8 mm (1.3 inches) in length and males only 3.4 mm (0.13 inches), for a size ratio of 9.6 (Kunter and Coddington, 2009).

2. For excellent comparative studies see Elgar (1991, 1998), Head (1995), Coddington et al. (1997), Prenter et al. (1999), Hormiga et al. (2000), and Foellmer and Moya-Laraño (2007).

3. Evolutionary biologists use the terms "orb-weaving" and "orb-web" to define a major evolutionary clade of spiders called the Orbiculariae that includes about a quarter of all known spider species. Early in the evolution of this clade, spiders evolved the archetypal orb web formed of radiating spokes of strong, dry silk joined by a spiral of stretchy capture silk that is coated with viscous adhesive secretions for trapping prey. This ancestral orb web is still found in many orbicularians, but it has been lost or greatly modified in several major lineages. As a result the majority of spiders included in the Orbiculariae, and hence considered orb-weavers in the evolutionary sense, do not spin orb webs at all. For example, sheet web spiders and cobweb spiders (including the infamous black widow and redback spiders) comprise almost half of all orbicularian species but spin sheet-like or irregular webs, as their names suggest. Perhaps for this reason most spider taxonomists, including those who write field guides and check lists, restrict the term "orb-weaver" to spiders in a single family of orbicularians, the Araneidae. This is a large, diverse family with over 160 genera and several thousand species, the majority of which do spin orb webs. However, some of the most well-known orb-web builders, the golden orb-weavers (including *Nephila* sp.), famous for both the large size of adult females (often over 5 cm long) and extreme sexual dimorphism, are classified in a different family altogether, the Nephilidae. Although all of this is indeed confusing, the case for *Argiope aurantia* is straightforward. The species is classified in the family Araneidae within the orbicularian clade, and juveniles and adult females spin classic orb webs with sticky capture spirals. Thus, they are orb-weavers by any criterion.

4. Elgar (1991) and Hormiga et al. (2000).

5. These estimates of average size come from Elgar (1991) and Hormiga et al. (2000). Other estimates for *Argiope aurantia* suggest that both sexes tend to be somewhat larger than this. For example, Howell and Jenkins (2004) give ranges of 19–28 mm and 5–8 mm for males and females, respectively.

6. Barth (2002).

7. Eberhard and Huber (2010) provide an excellent description and discussion of the structure and function of pedipalps in male spiders and how

they interact with the epigyna of females. See Foellmer (2008) for detailed information and photographs of pedipalps and embolus caps of male *Argiope aurantia*.

8. Hammond (2002) and Howell and Jenkins (2004).

9. Lockley and Young (1993).

10. Foellmer and Fairbairn (2004).

11. Foelix (1996).

12. Howell and Ellander (1984) and McReynolds (2000).

13. A strong positive correlation between female fecundity and body size (length or width) is a general rule for spiders (Head, 1995; Vollrath, 1998; Prenter et al., 1999; Foellmer and Moya-Laraño, 2007). For *A. aurantia*, the correlation between female prosoma width and number of eggs in the first egg sac is 0.53 under laboratory conditions (Matthias Foellmer, unpublished data). Lifetime fecundity in the laboratory is most strongly influenced by number of egg sacs, but prosoma width still has a highly significant effect (i.e., $p > 0.001$).

14. Howell and Ellender (1984), Foellmer and Fairbairn (2005a), and Matthias Foellmer (unpublished data).

15. Foellmer and Fairbairn (2003) and Foellmer (2008).

16. Gaskett (2007).

17. Foellmer and Fairbairn (2005b).

18. Gertsch (1979) and Foelix (1996).

19. For information about the types of prey captured and how females capture and subdue their prey, see Harwood (1974), Nyffeler et al. (1987), Foelix (1996), McReynolds (2000), Howell and Jenkins (2004), and Walter et al. (2008). The type of prey captured depends on habitat, season, and latitude; and within a given population, larger spiders capture larger prey.

20. Inkpen and Foellmer (2010).

21. Huber (2005).

22. Walter and Elgar (2012) review the adaptive significance of the conspicuous silk decorations on *Argiope* webs, and I am extrapolating from their conclusions. Although there is also evidence that web decorations enhance prey capture for some spider species by attracting insects, this has been shown not to be true for *A. aurantia*, and it would not likely apply to body colors.

23. My description of a male mating with a newly molted female (opportunistic mating) is taken from Robinson and Robinson (1980), Foellmer and Fairbairn (2003, 2005b), and Foellmer (2008).

24. Eberhard and Huber (2010).

25. This happens in more than 93 percent of first insertions (Foellmer, 2008).

26. Foellmer and Fairbairn (2003).

27. Male self-sacrifice during or immediately following insertion of their second pedipalp has evolved independently in at least six separate lineages of orb-weaving spiders (Miller, 2007). For excellent discussions of the adaptive significance of this behavior and of sexual cannibalism see Fromhage et al. (2005), Huber (2005), Millar (2007), Nessler et al. (2009), Wilder et al. (2009), and Wilder and Rypstra (2010).

28. See Foellmer (2008) for evidence that the embolus caps function as mating plugs in *A. aurantia* and a discussion of the evidence in other spiders, including *A. bruennichi* and *A. trifasciata*.

29. My description of matings with mature females comes from Foellmer and Fairbairn (2004).

30. Elgar et al. (2000), Schneider et al. (2006), and Herberstein et al. (2011).

31. Mortality during mate search has been estimated to be 86–89 percent for male redback spiders, *Latrodectus hasselti* (Andrade, 2003), 90 percent for golden silk spiders, *Nephila clavipes* (Vollrath and Parker, 1992), and 76 percent for golden orb-web spiders, *N. plumipes* (Kasumovic et al., 2007).

32. Moya-Laraño et al. (2008, 2009) and Prenter et al. (2010, 2012).

33. Corcobado et al. (2010).

34. Foellmer and Fairbairn (2005a) found that selection during mate search favored male *Argiope aurantia* with smaller prosomas and longer third legs in one population. Attempts to measure the effects of male size on success during roving in natural populations of other highly dimorphic spiders have yielded inconclusive results, partly because different authors use different indices of body size (sometimes using leg length as their measure of overall size) and have not measured leg length relative to body size. See Meraz et al. (2012) for a review of this literature.

35. Blanckenhorn (2000).

36. Foellmer and Fairbairn (2005a).

37. This size advantage does not apply to abdomen size, which varies greatly with fat reserves and so relates more to condition than to skeletal size (Foellmer and Fairbairn, 2005b).

38. In this final scramble, the male that happens to be closest to the female when she molts would seem to have the advantage. Surprisingly, males do not

seem to defend this position during the waiting period, and larger males are not more likely than smaller males to be there at the crucial time (Foellmer and Fairbairn, 2005a). This contrasts with golden silk spiders (*Nephila* sp.), where males do defend the hub, and larger males have the advantage (Elgar and Farley, 1996).

39. Researchers working on other orb-weaving spiders have faced a similar conundrum. In most species male size is either not important in mating interactions or larger males have the advantage. Where small males have an advantage it is in avoidance of female cannibalism during courtship and mating, whereas male-male scramble competition at the hub tends to favor larger males (Elgar, 1991; Elgar and Fahey, 1996). For examples of orb-weaving species where mating and fertilization success is not related to male size see Uhl and Vollrath (1998), Elgar et al. (2000), and Schneider and Elgar (2001). For examples in which larger males had an advantage in mating interactions, see Christenson and Goist (1979), Elgar and Nash (1988), Arnqvist and Henricksson (1997), and Johnson (2001). Golden silk spiders, *Nephila edulis,* are an exception to this (Schneider et al., 2000). Small and large males of this species employ different mating tactics, and small males are more successful.

40. Vollrath (1998).

41. Fromhage et al. (2005, 2008).

42. In spiders the term "female gigantism" may be more appropriate because in most groups, including *Argiope*, the evolutionary process seems to have been one of increasing female size rather than decreasing male size (Prenter et al., 1999; Hormiga et al., 2000).

43. See appendix B and the detailed information by class archived in datadryad.org (Fairbairn, 2013).

Chapter 7: Blanket Octopuses

1. Prenter et al. (1999), Hormiga et al. (2000), and Foellmer and Moya-Laranño (2007).

2. Ghiselin (1974), Andersson (1994), and Vollrath (1998).

3. For detailed descriptions and classification of blanket octopuses see Thomas (1977), O'Shea (1999), and Mangold et al. (2010).

4. For general and not overly technical descriptions of octopods, see Nesis (1987) and Hanlon and Messenger (1996). More details concerning the life cycles and ecology of individual species can be found in Boyle (1987).

5. A few species lack a radula.

6. See Hanlon and Messenger (1996) for more information about general cephalopod behavior and intelligence.

7. The Argonauts were sailors on the ship Argo that carried Jason on his quest to find the Golden Fleece.

8. The superfamily Argonautoidea contains four families, each containing a single genus. The families, genera, and number of species are the blanket octopuses, Tremoctopodidae (*Tremoctopus,* four species); the paper nautiluses, Argonautidae (*Argonauta,* four species); Ocythoidae (*Ocythoe,* one species); and Alloposidae (*Haliphron,* one species). Good descriptions of the species in each genus can be found in Nesis (1987), Young and Vecchione (2008), Mangold et al. (2010a,b), and Young (2010).

9. The data on octopus size come from Nesis (1987), Rutledge (2000), Norman et al. (2002), O'Shea (2010), and Young (2010). The two tiny octopus species are *Octopus micropyrsus* and *nanus,* the former being the smallest of living octopods. The *Argonautus* species with similarly sized males are *nodosa* and *argo.*

10. Norman et al. (2002).

11. For descriptions of the use of *Physalia* tentacles and general foraging behavior in blanket octopuses, see Jones (1963), Thomas (1977), Hanlon and Messenger (1996), Norman et al. (2002).

12. Thomas (1977) describes the allometric growth of females.

13. See Thomas (1977) for illustrated descriptions of growth and development in males and females.

14. Thomas (1977), Hanlon and Messenger (1996), and Andersen et al. (2002).

15. For descriptions of the life histories and reproductive behavior of female blanket octopuses, see Boyle (1987), Hanlon and Messenger (1996), and Laptikhowsky and Salman (2003). Young (1996) provides excellent photographs and a description of blanket octopus egg clusters and embryos.

16. Hanlon and Messenger (1996) and Rocha et al. (2001).

17. See Andersen et al. (2002) for a discussion of senescence and O'Dor and Wells (1987) for a discussion of energy metabolism in cephalopods. The effect of size on survival during fasting is related to the allometry of energy storage versus energy usage. Larger animals use less energy per gram of tissue, but they require more energy overall simply because they are larger. If the amount of energy that can be stored prior to fasting scales proportionally with size, smaller animals are at a disadvantage because they use up their reserves faster. Larger animals will

live longer. The effect is magnified if energy reserves scale positively with body size (larger animals have proportionally more energy reserves).

18. Hanlon and Messenger (1996).

19. See Blanckenhorn et al. (1995) and Blanckenhorn (2000, 2005) for discussions of the advantages of small size in males that must search for mates. If the activities associated with finding food and eating detract from mate searching, and foraging success does not increase disproportionately with body size, small size is favored because smaller males are able to devote more time to mate search and less to foraging than larger males. Roff (1991) also argues that small size is favored in ectothermic animals that depend on passive transport by wind or current for long distance movement. His argument would apply to male *Tremoctopus* species.

20. According to Wells and Wells (1977).

21. Nesis (1987).

22. Thomas (1977).

Chapter 8: Giant Seadevils

1. The order Lophiiformes refers to all anglerfishes. The deep-sea anglerfishes are in the suborder Ceratioidei and this includes eleven families, ten of which have the common name "seadevil." The other family (Himantolophidae) has the common name "football fish."

2. See Pietsch (2009) for a complete discussion of the phylogeny and classification of all the ceratioid anglerfishes, species descriptions and detailed information about their distribution, ecology, and life history. The majority of information in my description comes from this comprehensive text. A much condensed but easily accessed summary of this information can be found online in Pietsch and Kenaley (2007).

3. Bertelsen (1951) and Pietsch (2005, 2009).

4. Bertelsen (1951).

5. Pietsch (2009).

6. Females have virtually no vision and only very small olfactory organs, so they probably detect prey by the pressure waves generated as they approach and by touch (Pietsch, 2009).

7. Low encounter rates between sexes is a general problem for bathypelagic fishes and may be a major limitation on net reproduction, independent of

per-capita resource availability. See Baird and Jumper (1995) for a discussion of this and an analysis of encounter probabilities as a function of density and detection distance. Reduction in individual reproductive success (Darwinian fitness) as a consequence of low population densities is a general phenomenon in ecology and is known as the "Allee effect." Fisheries biologists use the term "depensation" for the same phenomenon.

8. Albret Eide Parr (1930) was the first to recognize that male ceratioid fishes, formerly classified in their own family, Aceratiidae, were free-living males belonging to the same families as female ceratioids. Danish researcher Erik Bertelsen (1951) was the first to be able to assign free-living males to the correct ceratioid genera and species.

9. Not all ceratioid anglerfishes have parasitic males (Pietsch, 2009). Obligate sexual parasitism is found in four of the eleven families, including the Ceratiidae, the family to which *Ceratias* belongs, and sexual parasitism is facultative in a fifth family. In the other families males only attach themselves to females temporarily and do not become parasitic.

10. The extreme sexual size dimorphism claimed for *C. holboelli* is based on the sizes of free-living males compared to females. Based on the records of parasitic males and their mates in Pietsch (2009), the sexual size ratio for standard lengths averages 9.1 (range 6.5–16.0, $n = 11$), and for total length averages 10.4 (range 6.4–13.8, $n = 5$).

11. Female monkfish, which are commercially fished, shallow-water anglerfishes, do not become sexually mature until they are eight to eleven years old (http://www.fishonline.org/fish/monkfish-anglerfish-149, accessed September 24, 2011). It seems likely that seadevils take at least this long because they grow to similar sizes (Landa et al., 2001) but in much colder waters and with much lower densities of prey.

12. The data in Pietsch (2009) show that females metamorphose from the larval to the juvenile stage at standard lengths ranging from only 1.7 to 2.1 cm (0.67–0.83 inches) By comparison, the average standard length of females with an attached male is 68 cm (26.8 inches) and the smallest female with an attached male has a standard length of 56 cm (22.0 inches).

13. Pietsch and Grobecker (1987).

14. The strategy of releasing eggs in a gelatinous mass is common to all anglerfishes (Lophiiformes) and so probably evolved in shallow water and in species without dwarf males. The presence of this method of spawning may have facilitated colonization of the deep, open ocean by the ancestors of ceratioid anglerfishes.

15. According to Bertelsen (1951) and Pietsch (2009) this is a general rule for ceratioid anglerfishes, at least in the North Atlantic where most specimens have been caught.

16. Roff (1992) and Lowerre-Barbieri (2009).

17. The smallest female recorded with an attached male had a standard length of 56 cm (22.0 inches) and her male had a standard length of 3.5 cm (1.4 inches). Both are the smallest in the data set (Pietsch, 2009). The correlation between male and female standard lengths is 0.523, $p < 0.05$ ($n = 10$, excluding a female that had two attached males).

18. Crozier (1989) found a maximum age of sixteen years for a *Lophius piscatoriuis*. Landa et al. (2001) estimated the maximum age of female *Lophius piscatorius* to be twenty-two years, and that of female *L. budegassa* to be nineteen years. The website of the Marine Conservation Society of the UK (http://www.fish online.org/fish/monkfish-anglerfish-149, accessed September 24, 2011) gives a maximum lifespan of twenty-four years for the two species combined.

Chapter 9: Bone-Eating Worms

1. Examples of groups where only sperm are released into the water and fertilization occurs inside the mother's body include most sponges, mussels, oysters, and moss animals. Both sperm and eggs are released into the water, and fertilization occurs in the open in most clams, corals, lampshells, and sea lilies.

2. The description of this initial discovery and two subsequent visits to the whale carcass can be found in Goffredi et al. (2004). The carcass was initially described as being at a depth of 2891 m (9,483 feet), but in later papers this is given as 2,893 m (9,493 feet) (Vrijenhoek et al., 2009; Lundsten et al., 2010).

3. Dr. Robert Vrijenhoek is a Senior Scientist at MBARI. Dr. Shana Goffredi is now an Assistant Professor of Biology at Occidental College in Los Angeles, California, and Dr. Greg Rouse is now Curator of the benthic invertebrate collection and Director of the Marine Invertebrate Phylogenetics Laboratory at the Scripps Institute of Oceanography, San Diego.

4. Rouse et al. (2004).

5. Hilário et al. (2011).

6. The two new *Osedax* species they describe were thought initially to be only one species, called "annelid A" in Goffredi et al. (2004).

7. Lundsten et al. (2010).

8. Glover et al. (2004), Fujikura et al. (2006), and Vrijenhoek et al. (2009).

9. Rouse et al. (2011).

10. Of the seventeen species described to date, only one is topped by a pale, spiral-shaped structure in lieu of palps. The rest have four feathery palps in various shades of pink or red (Vrijenhoek et al., 2009).

11. See Katz et al. (2011) for a detailed description of the *Osedax* trophosome and a comparison with the trophosomes of other siboglinid worms.

12. These are marine bacteria from the order Oceanospirillales, a group of bacteria known to be able to degrade complex organic compounds.

13. See Goffredi et al. (2005, 2007) and Verna et al. (2010) for more information about the symbiotic bacteria and their role in *Osedax* nutrition.

14. Braby et al. (2007), Rouse et al. (2008, 2009), Vrijenhoek et al. (2008, 2009), and Lundsten et al. (2010).

15. My descriptions are based mainly on *Osedax rubiplumus, frankpressi,* and *roseus,* simply because they have been described in the most detail. The first two have only been found on the deeper carcasses (at depths of 1,820 m and 2,893 m), whereas *roseus* occurs on the somewhat shallower carcasses (633 m, 1,018 m, and 1,820 m). Both *rubiplumus* and *roseus* seem to colonize whale-falls quickly and achieve their maximum densities within the first six months, whereas *frankpressi* arrives slightly later, increases in abundance more slowly, and takes a year or more to reach maximum densities. These differences are minor in comparison to the overall similarities in ecology and life history among the three species.

16. Rouse et al. (2009) found that in the laboratory, at temperatures of 4°–6°C, the larvae of several *Osedax* species become free-swimming within 48 hours. However, development may be slower in nature because of lower oxygen content of the water.

17. Rouse et al. (2009).

18. Natural carcasses in Monterey Canyon are spaced 5 to 15 km (approximately 3 to 9 miles) apart, although carcasses may have been more abundant before the days of whaling (Rouse et al., 2008; Vrijenhoek et al., 2008).

19. Rouse et al. (2008).

20. Lundsten et al. (2010).

21. For descriptions of the life history of female *Osedax* see Rouse et al. (2008, 2009) and Vrijenhoek et al. (2008).

22. For descriptions and photographs of larval and male *Osedax*, see Rouse et al. (2004, 2008, 2009), Vrijenhoek et al. (2008), and Worsaae and Rouse (2009).

This life history in which sexually mature individuals retain their larval morphology goes by the scientific name paedomorphosis.

23. A conservative estimate of average female lifespan would be six to nine months, based on the time series of female size distributions of *O. roseus* in Rouse et al. (2008).

24. Deep-water species such as *roseus, frankpressi,* and *rubiplumus* cannot be observed spawning in laboratory tanks because of the high pressure under which they normally live. However, Rouse and his colleagues (Rouse et al., 2009) were able to collect and observe spawning in a shallower-water species with the unofficial name "orange collar" that they had collected from a whale deployed at 633 m (2,077 feet).

25. Rouse et al. (2004, 2008) and Vrijenhoek et al. (2008).

26. In the samples collected by Rouse et al. (2008) and Vrijenhoek et al. (2008), 18 percent of *rubiplumus* and 16 percent of *roseus* females had acquired males but did not yet have eggs in their oviducts.

27. See figures 1 and 2 in Rouse et al. (2008). The Pearson correlation coefficient, r, between female size (measured as preserved crown-trunk length) and number of males was 0.65 ($n = 103, p < 0.001$) for *O. roseus* in the nine-month sample (Rouse et al. 2008). By comparison, the correlation between female trunk width and number of males was reported to be 0.80. 0.89, and 0.90 for populations of *O. rubiplumus* on whales 1820 and 2893 ($p < 0.01$, Rouse et al., 2004; and $p < 0.001$, Vrijenhoek et al., 2008). In *O. rubiplumus,* females with eggs in their oviducts were significantly larger than those without eggs, and females with harems were significantly larger than those without (Vrijenhoek et al., 2008).

28. For more information about sex determination and reproductive strategies in green spoon worms, see Jaccarini et al. (1983) and Berec et al. (2005). Rouse et al. (2004, 2008, 2009) and Vrijenhoek et al. (2008) discuss the applicability of this to *Osedax*.

29. Rouse et al. (2008), Vrijenhoek et al. (2008), and Metaxas and Kelly (2010).

Chapter 10: Shell-Burrowing Barnacles

1. Barnacles comprise the arthropod infraclass Cirripedia. For ease of illustration I describe the typical acorn and gooseneck barnacles that are found

attached to hard substrates. This is the "mainstream" mode of life for barnacles (Anderson, 1994), but the Cirripedia also includes species that burrow into substrates, live on other organisms commensally or as exo- or endoparasites, or remain free-swimming in the plankton.

2. Kelly and Sanford (2010).

3. This general description of barnacles comes mainly from Anderson (1994).

4. See Yamaguchi et al. (2006) and Kelly and Sanford (2010) for recent analyses of reproductive strategies in barnacles and specifically, predictions concerning the conditions that favor pure hermaphroditism, androdioecy (complemental males plus hermaphrodites), or dioecy with dwarf males.

5. Darwin wrote two long monographs on living barnacles that still stand as excellent references today (Darwin, 1851, 1854). He also discusses barnacle mating systems and sexual dimorphism in *The Origin of Species* (Darwin, 1859). The quotation I give can be found on page 421 of the first edition of *The Origin* (1859) published by Penguin Classics in 1985 and on page 420 of the sixth edition (1872) published by Mentor in 1986.

6. Thoracica and Acrothoracica are usually given the taxonomic rank of superorders (Anderson, 1994; Tree of Life, 2009), although older publications refer to them as orders (e.g., Tomlinson, 1969a). Acorn barnacles are in the order Sessilia, and gooseneck barnacles are in the order Pedunculata within the Thoracica. The burrowing barnacle *Trypetesa lampas* is in the order Apygophora within the superorder Acrothoracica.

7. The descriptions in this paragraph are taken from Tomlinson (1969a,b) and Andersen (1994).

8. My detailed description of *Trypetesa lampas* comes primarily from Tomlinson (1969a), White (1969), Gotelli and Spivey (1992), Andersen (1994), and Williams et al. (2011). Statistics for its distribution and prevalence in gastropod shells come from White (1969), McDermott (2001), and Reiss et al. (2003).

9. Whelks and conchs are marine mollusks in the class Gastropoda.

10. Gotelli and Spivey (1992).

11. Most crustaceans, including thoracican barnacles, have six naupliar stages, and many feed during the naupliar phase of the life cycle. The fourth-stage nauplius of *T. lampas* is developmentally equivalent to the sixth-stage nauplii of other crustaceans.

12. Most burrowing barnacles brood the larvae to the cyprid stage, but the nauplii of *T. lampas* are released from the mantle cavity (Gotelli and Spivey, 1992).

13. White (1970).

14. Gotelli and Spivey (1992) did not determine if all of the 207 females they sampled were reproductively mature but seem to have assumed that they all were.

15. Andersen (1994).

16. See Ghiselin (1974), Coddington (1997), Vollrath (1998), and Blanckenhorn (2000, 2005) for discussions of the hypothesis that small males are more efficient in mate searching as well as other hypotheses for the existence of dwarf males.

17. One could argue that males could mature early without also being so small if they simply grew faster than females, and so they need not be dwarfed. Differences between the sexes in rates of growth are indeed quite common, as we have seen. However, the typical pattern is that the larger sex grows faster than the smaller, not the reverse, and there are no examples in which growth rate differences alone can produce the extreme differences in age at maturity found in species with dwarf males. Blanckenhorn et al. (2007) evaluated the relationship between sexual size dimorphism and differences between the sexes in development time and growth rate in seasonally breeding insects and spiders where males were the smaller sex. They found that, except in species where earlier maturation of males was directly favored, size differences between the sexes tend to be caused by differences between the sexes in growth rates rather than differences in development time (i.e., time to reach maturation), so the sexes matured at the same time but at different sizes. The pattern was that females grew faster and matured at a larger size than males. Extreme size dimorphisms were associated with slower growth and earlier maturation of males combined with faster growth and later maturation of females.

18. See Yamaguchi et al. (2007), Urano et al. (2009), and Kelly and Sanford (2010). Based on simulation models and life history theory, they conclude that males forgo growth for reproduction in environments where group sizes are small and lack of food severely limits growth. At the extreme, males do not grow at all, do not develop digestive systems, and devote all of their resources to reproduction.

Chapter 11: The Diversity of Sexual Differences

1. Dioecy occurs in sixty-seven classes distributed among twenty phyla, plus six small phyla with no recognized classes. For simplicity I treat the phyla

with no class subdivisions as though they were each a single class, for a total of seventy-three "classes." Appendix B shows my scores for the prevalence and magnitude of sexual dimorphism within each phylum and also lists the types of dimorphisms found. More extensive descriptions of the sexual differences in each of the seventy-three classes can be found in the online supporting information archived in the Dryad repository at http://dx.doi.org/10.5061/dryad .n48cm (Fairbairn, 2013).

2. The percentages and proportions given are based on the presence or absence of each type of dimorphism within each animal class. This is likely to be a poor reflection of prevalence in terms of numbers of species or numbers of individuals showing the dimorphism because the number of species per class ranges from >1 million in the class Insecta to only 1 in the phylum Placozoa. However, there are simply too many species (>1.4 million) to perform this type of survey for every species known. Even if this could have been done, working at the species level would inordinately bias the conclusions toward the patterns found in a few very large classes, such as the Insecta, the Arachnida (74,000 species), the Malacostraca (27,000 species), and the Actinopterygii (24,000 species). This is a conundrum. I settled on taxonomic class as my unit of study because this gave me a workable number of groups to characterize while at the same time it captured the sweep of animal diversity in terms of different body plans and developmental programs. For readers interested in prevalence at the level of species, appendix B gives a rough rank of the proportion of species with detectable sexual dimorphism in each phylum.

3. Visible gonadal dimorphisms occur in all classes of cnidarians and ctenophorans, but in only three phyla and six classes of lophotrochozoans, and two phyla and three classes of deuterostomes.

4. The male genital exit has been modified to serve as an intromittent organ during copulation in thirty classes and twelve phyla.

5. Examples of genital dimorphisms in free-spawning species that do not have genital contact include sea urchins, sea cucumbers, brittle stars, and lampreys.

6. Females average larger than males in reptiles (especially turtles and snakes) and in the majority of amphibians and fishes, including bony fishes, sharks, and rays.

7. Among arthropods the only exception is the class Malacostraca, which contains crustaceans such as lobsters, crabs, and shrimp, where males average larger than females.

8. Females are larger in most spiny-headed worms (Acanthocephala), segmented worms (Annelida), spoon worms (Echiura), roundworms (Nemata), horsehair worms (Nematomorpha), ribbon worms (Nemertea), velvet worms (Onychophora), flatworms (Platyhelminthes), and pterobranchs (Hemichordata).

9. This trend is quite robust across the animal kingdom, appearing in the acanthocephalans, cycliophorans, nematodes, nemerteans, maxillopod crustaceans, mesozoans, and cephalopod and gastropod mollusks. For comprehensive reviews of sexual size dimorphisms in many parasitic taxa, see Poulin (1996, 1997) and Poulin and Morand (2000).

10. This estimate for Leopard seals comes from Alexander et al. (1979). Weckerly (1998) gives a mass ratio of only 1.13. Leopard seals just miss the distinction of being the mammal with the largest female-larger dimorphism. That title goes to a tiny bat, the little white-shouldered bat, in which the tiny females, weighing only 12 g, average 66 percent heavier than males (Ralls, 1976, and http://animaldiversity.ummz.umich.edu/site/accounts/information/Ametrida _centurio.html. Accessed September 1, 2009).

11. The greatest male-larger size dimorphism among all birds is found in the great bustard, which we met in chapter 4, with males as much as 3.14 times heavier than females. Muscovy ducks rank second with a male:female mass ratio of 2.40.

12. See Parker (1992), Roff (1992), Fairbairn (1997), Taborsky (2001), Lindenfors et al. (2007), and Székely et al. (2007) for extensive evaluations of the hypotheses presented in this section.

13. Larger males tend to win contests because size conveys strength and inertia—and also because overall size is often associated with larger weapons (such as elongated canines, tusks, horns, or antlers in mammals) as well as with greater age and experience. See Reiss (1989) for an excellent explanation of why large size becomes increasingly advantageous in physical contests as animals increase in size.

14. See Ruckstuhl and Neuhaus (2005) for excellent examples of this.

15. In birds, smaller males may have the advantage in species where competition and female choice are based on aerial displays (Colwell, 2000 and Székely, 2004).

16. For reviews of the evidence for this see Andersen (1994), Fairbairn (1997), Blanckenhorn (2005), and Fairbairn et al. (2007).

17. Ralls (1976), Wedell et al. (2002), and Clutton-Brock (2002).

18. Fairbairn (1988), Arak (1888), Preziosi and Fairbairn (2000), and Foellmer and Fairbairn (2005a).

19. Males average larger than females only in mammals and birds, two Arthropod classes (Malacostraca and Ostracoda), the phylum Entoprocta, and the molluskan class Polyplacophora.

20. The scenario described has been proposed to explain male dwarfism in a number of different free-living animal lineages, including deep-ocean fishes, open-ocean cephalopods (octopods and squid), various groups of bottom-dwelling marine worms, and orb-web spiders, as well as in various lineages of parasitic and commensal species. See Vollrath (1998) and Blanckenhorn (2000) for general discussions.

21. "Wormy" exceptions are flukes and flatworms (Platyheminthes), where females are thinner than males.

22. For examples from insects, see Emlen and Nijhout (2000).

23. Brana (1996), King (2009), and Seifan et al. (2009).

24. Examples of copulatory organs include the specialized pedipalps of male spiders, modified pleopods in malacrostracan crustaceans such as crabs and lobsters, and modified fifth legs in seed shrimp (Ostracoda).

25. Roff (1990).

26. Throughout this section I am using coloration to refer to visually detectable patterns in the wavelengths of light reflected and absorbed by the external covering (skin, hair, feathers, scales, etc.) of an animal. In that sense, black (all color absorbed) and white (all color reflected) are both considered "colors." In most cases these patterns are caused by pigmentation in the integument, but perceived coloration can also be caused by physical properties of the integument. The flash of red or purple from a male hummingbird's chin and the bright green of a male mallard's head are familiar examples of this.

27. See Badyaev and Hill (2003) for an excellent review of sexual dichromatism in birds.

28. See Sinervo and Lively (1996), Wiens et al. (1999), and Seifan et al. (2009) for studies of sexual dichromatism in lizards and Pianka and Vitt (2003) for less technical descriptions and photographs.

29. Mimicry in milk snakes is nicely described in Pfennig et al. (2001).

30. See Endler and Basolo (1998) for a review of the evolution and adaptive significance of visual signals.

31. This hypothesis has been particularly well studied in birds. Excellent discussions of the evolution, proximate causes, distribution, and adaptive

significance of color dichromatisms in birds can be found in Owens and Short (1995), Kimball and Ligon (1999), and Badyaev and Hill (2003). Plumage coloration has been shown to be condition-dependent in some sexually dichromatic species, but the colors of beaks, wattles, and skin patches are often more obvious and reliable indicators of male condition (Owens and Short, 1995). Bennett et al. (1994) discuss how visual perception systems in birds relate to the evolution of sexual dichromatisms, and Endler and Mielke (2005) provide a more technical treatment of this subject with explicit comparisons of human and bird color perception. For more general reviews across taxa, see Halliday (1980), Zuk (1991), Chronin (1991), Ridley (1993), Andersson (1994), and Cotton et al. (2004).

32. By selecting healthier mates, females avoid contact with males in poor condition that may transmit parasites or pathogens. Such discrimination may also benefit females indirectly by identifying males that carry good genes for growth and survival and for sexually attractive traits. By mating with these males, females increase the chances that their offspring will also be genetically well endowed and so will successfully survive and reproduce and that their sons will themselves be attractive to females and so have high mating success. The evidence for these two hypotheses, called "good genes" and "sexy son," respectively, is reviewed in Cotton et al. (2004) and Prokop et al. (2012).

33. See Gray (1996) and Cotton et al. (2004) for evidence of the general condition-dependence of sexually selected traits.

34. For dragonflies see Moore (1990); for butterflies see Rutowski (2003), Wicklund (2003), and Costanzo and Monteiro (2007); for red-winged blackbirds see Smith (1972); for side-blotched lizards see Sinervo and Lively (1996); and for sticklebacks see Kraak et al. (1999).

35. For examples see Endler and Houde (1995), Wiens et al. (1999), Badyaev and Hill (2003), ffrench-Constant and Kock (2003), Croft et al. (2004), Millar et al. (2006), and Kunte (2008).

36. This uncommon pattern of reverse sexual dichromatism occurs in some spiders (including our yellow garden spiders), crabs, shrimp, and hemichordates, and in a few species of birds, lizards, fishes, insects, snails, chitons, ribbon worms, and box jellies.

37. For an excellent review of the theory and evidence for male mate choice, see Bonduriansky (2001). He gives a few examples of male preferences based on female pigmentation in insects. For examples of lizards see LeBas and Marshall (2000) and Weiss (2006); for birds see Roulin (1999), Amundsen et al. (1997),

and Amundsen (2000); for crabs see Williams (2003); and for fishes see Berglund and Rosenqvist (2001) and Houde (2001).

38. For examples of female-female competition see Amundsen (2000), Berglund and Rosenqvist (2001), and Hegyi et al. (2007), and for a discussion of the conditions under which such competition is likely to occur, see Berglund et al. (1993).

39. See West-Eberhard (1983) and Amundsen (2000) for the advantages of bright coloration in females. Also see Arnqvist and Nilsson (2000) and Zeh and Zeh (2003) for excellent discussions of the costs and benefits of multiple mating in females. Both papers discuss the mechanisms by which multiple mating by females could increase female fitness. Sperm limitation (inability to obtain sufficient sperm to ensure full fertility) is unlikely to be a problem for females in most species with internal fertilization, and most females obtain more sperm than necessary to fertilize all of their eggs. For that reason it is likely that females in most species do not compete simply to obtain more sperm. It is more likely that they are competing for resources that males provide that enable them to produce more eggs or for sperm with higher genetic quality. Sperm limitation may be quite common in sessile, free-spawning, externally fertilizing species that exist in low densities (Levitan and Petersen, 1995, and Levitan, 1996), and as we have seen, even in some species with internal fertilization that exist in very low densities, but these are not species in which we would expect sexual selection on color.

40. Berglund and Rosenqvist (2001) and Badyaev and Hill (2003).

41. Bauer (1981) and Jormalainen et al. (1995).

42. Detto et al. (2006) and Kunte (2008).

43. For a comprehensive review of what is known about color vision and the use of color for both signaling and camouflage in both vertebrate and invertebrate animals, see Kelber et al. (2003).

44. Land and Nilsson (2004).

45. Whereas most mammals other than primates have two cone types, whales and seals have only one cone and see only green light (Peichi et al., 2001).

46. Vertebrates other than mammals typically have four cone types, but snakes, crocodiles, and geckos have only three. Many birds have five cone types and also have oil droplets in their cones that enable them to discern finer gradations of color than we can. Flies and butterflies have at least five cone types, most dragonflies have four, bees have three, and other insects tend to have two or three. Many spiders have only two cone types, but jumping spiders typically

have four. Most crustaceans have three cone types, but mantis shrimp, which are famous for their incredible colors, have twelve or more. Most octopus and squid have only one spectral-sensitive receptor and so see only one color, which seems surprising given their famous ability to rapidly change color and pattern.

47. Garm et al. (2007).

48. Endler (1992, 1993) and Gomez and Théry (2007).

49. The exceptions are the arthropod classes Pauropoda and Symphyla, the mollusk class Monoplacophora, and the poriferan class Calcarea. These are minor classes totaling fewer than 1,100 species.

Chapter 12: Concluding Remarks

1. In primates sexual dimorphism is strongly associated with the intensity of sexual selection on males through both competition among males to monopolize matings with females and female choice of mates, and sexual selection also acts on female primates through both competition among females and male-mate choice (Geary, 1998; Lindenfors and Tullberg, 1998; Lindenfors, 2002; Plavcan, 2004; and chapters 3–6 in Kappeler and van Schaik, 2004). There is abundant evidence that sexual selection continues to operate on both sexes in human populations and targets a variety of traits including body size, face shape, and body shape (Geary, 1998; Johnson and Tassinary, 2007; Weston et al., 2007; Courtiol et al., 2010).

2. When males are the larger sex it is common for sexual size dimorphism to be positively correlated with body size, so that males and females differ more in size in larger species and less in size in smaller species (Fairbairn, 1997; Fairbairn et al., 2007). This pattern is strong in the clade to which we belong, the haplorhine primates (Lindenfors and Tullberg, 1998; Lindenfors, 2002), but our sexual size dimorphism is less than would be expected for a primate of our size. Fossil evidence supports the conclusion that sexual dimorphism for body size decreased as we evolved from our early hominid ancestors (Ruff, 2002; Skinner and Wood, 2006).

3. Authors of popular books and articles on sex and gender in humans frequently assume that various cognitive, behavioral, and psychological traits attributed to men and women in our societies are inevitable consequences of being male or female. In this literature traits such as high activity levels, aggressiveness, and proclivity for promiscuity and mathematics are frequently attributed

to males, whereas enhanced verbal skills and more passive, coy, and nurturing behavior are attributed to females, as though these were fixed, dichotomous differences between the sexes. To the contrary, although many of these differences do have a genetic basis, there is abundant evidence that these traits are also heavily influenced by developmental environments and that their distributions overlap greatly between the sexes. Even morphological traits such as height and proportion of body fat are subject to such cultural influences through different allocation of diet, physical labor, and exercise to boys and girls. Among the many examples of typological thinking about sex differences, see Ridley (1993), Baron-Cohen (2003), Brizendine (2006), Gonzales (2005), Quirk (2006), and Benjamin (2012). Among the many excellent critiques see Fausto-Sterling et al. (1997), Valian (1998), Zuk (2002), Guiso et al. (2008), Fine (2010), Jordan-Young (2010), and Eliot (2010). For evidence of genetic differences between the sexes, see Skuse (2006), Weiss et al. (2006), and Pan et al. (2007).

4. Adult males are taller on average than adult females in all human populations. The average difference in height is 11.4 cm and ranges from 6.8 cm in the Mtubi of Africa to 17.3 cm in the Blackfoot Indians of North America (Holden and Mace, 1999). The data given in the text are for adults in the United States and come from the National Center for Health Statistics Public Health Service Bulletin No. 1000, Series 11, No. 14, 39 pp. (1966).

5. See note 3.

Appendix B: Summary of Sexual Dimorphisms

1. This classification is from the ITIS Catalogue of Life: 2008 Annual Checklist (Bisby et al., 2008).

2. A listing of this information for each class within each phylum, including more complete descriptions of sexual differences and data sources, is archived in the Dryad repository at http://dx.doi.org/10.5061/dryad.n48cm (Fairbairn, 2013).

GLOSSARY OF TECHNICAL TERMS

Term	Definition
animal	A member of the Kingdom Animalia; a multicellular, eukaryotic, heterotrophic organism that obtains nutrients from the external environment and digests them internally.
bathypelagic	*adj.* Living and feeding in the water column at depths between 1,000 m and 4,000 m, where no sunlight penetrates.
benthic	*adj.* Dwelling on, or relating to, the bottom of a body of water; living on or very close to the bottom of a lake, ocean, or other body of water.
bioluminescence	The production and emission of light by a living organism.
capital breeding	A reproductive strategy in which individuals do not feed while engaged in reproductive activities (e.g., courtship, mating, oviposition) and must use the energy gained from prereproductive foraging.
caruncle	A club-shaped appendage on the back of female ceratioid anglerfish in the family Ceratiidae that contains bacteria-filled photophores.
cilia	Plural of cilium, a microscopic, thread-like, or hair-like structure formed of microtubules and protruding from eukaryotic cells. Motile cilia beat in coordinated waves and serve as locomotory organelles for many types of larvae and small eukaryotes.
clade	An evolutionary lineage containing all descendants from a single common ancestor and forming a distinct branch on a phylogenetic (i.e., evolutionary) tree.

commensal   *adj.* Living close to, on, or within another organism and benefiting from this; the host organism is not affected by this relationship. If the host is harmed, the relationship is called parasitism. If the host benefits, it is called mutualism.

condition   When used to refer to animals, usually means some combination of general health (absence of disease or injury) and presence of energy reserves in the form of stored fat, carbohydrate, and protein. Often measured relative to the average body weight expected for a given measure of linear or skeletal size.

copulation   Physical contact of two sexual individuals resulting in deposition of the male gametes from one partner on or within the body of the other partner in proximity to the female gametes. Usually requires a specialized male organ for sperm transfer (the intromittent organ) and a specialized female receptacle.

dichromatism   A dimorphism in the color or pigment pattern caused by differences in the wavelengths of light reflected from the body surface.

dioecious   *adj.* Refers to a population or species in which male and female reproductive organs occur in separate individuals (i.e., the sexes are separate; individuals are gonochoric).

dioecy   The allocation of reproductive roles in a population such that all individuals are either male or female and possess only male or female reproductive organs (i.e., see dioecious).

diploid   *adj.* Refers to the chromosome complement of a cell. Diploid cells have two sets of chromosomes, one inherited from the father and one from the mother. Thus, a diploid cell possesses a pair of chromosomes of each type. (See haploid.)

diurnal   *adj.* Associated with the daylight portion of the twenty-four-hour day. Usually refers to organisms that are active primarily during the daylight hours.

egg raft

A buoyant mass consisting of fertilized eggs embedded in a sheath of mucoid, gelatinous material, produced by marine anglerfishes in the order Lophiiformes.

embolus

The heavily sclerotized (hardened) tip of the pedipalp of male spiders that serves as the intromittent organ.

epigenetic effects

Heritable changes in gene expression caused by mechanisms other than changes in the underlying DNA sequence.

epigynum

A slightly raised, sclerotized (hardened) plate on the underside of the abdomen of female spiders that contains the paired genital openings.

epipelagic

adj. Refers to the sunlit or euphotic waters of the ocean where photosynthetic activity is possible. Typically extends from the surface down to about 200 m.

esca

The fleshy lure or bait at the tip of the illicium of anglerfishes (Teleostei, Lophiiformes), including seadevils (Ceratioidei, Ceratiidae).

estrus

The portion of the mammalian reproductive cycle when females have maximum sexual receptivity and fertility; usually also the time of release of the eggs.

fecundity

The number of eggs or offspring produced by a female. Usually refers to the number per breeding event (i.e., per clutch, batch, or litter) unless specified as lifetime fecundity.

fledgling

A chick that is morphologically mature and able to fly.

fitness

An abstract concept that refers to the success of an individual in contributing genes to the gene pool of future generations.

fry

A juvenile fish that has absorbed its yolk sack and is capable of independent feeding.

gamete

A reproductive cell that contains one-half of the chromosome complement of the adult, diploid parent. In animals, gametes produced by males are sperm, and those produced by females are eggs.

gender	Gender is a grammatical classification of objects into male and female classes. It is often used colloquially as a synonym for sex, as in male and female sexes.
gonad	An organ whose primary function is to produce gametes; ovaries and testes in animals.
gonochoric	*adj.* The presence of reproductive organs of only one sex within an individual. A dioecious population consists only of gonochoric individuals.
haploid	*adj.* Refers to the chromosome complement of a cell. Haploid cells have only one chromosome of each type, whereas diploid cells possess a pair of chromosomes of each type.
hectocotylus	The modified arm of male cephalopods used as a copulatory organ.
hermaphrodite	An individual organism that is able to produce both male and female gametes in its lifetime.
illicium	The modified first dorsal spine in anglerfishes (Teleostei, Lophiiformes) that supports the lure or esca.
lek	A gathering of males for the purposes of competitive mating display to attract females, usually in a traditional location or arena. Females visit the lek to choose a mate and copulate but do not nest or obtain other resources on the leking grounds.
lordosis	In zoology, a sexually receptive posture adopted by some female mammals in which the spine is curved so that the pelvis tips forward to lift and expose the genital opening to facilitate mating.
mutualism	A close ecological relationship (symbiosis) between two different species in which both species benefit from the association.
natural selection	Change across generations in biological characteristics caused by consistent relationships between heritable trait values and fitness.
oviposition	Egg laying. To oviposit means to lay eggs.
palps	(1) Sensory appendages that arise from the mouth parts of many arthropods, including

crustaceans, insects, and spiders. (2) Tactile organs arising from the head or anterior end of certain annelids and mollusks. (3) The distal segments of the pedipalps of spiders, excluding the coxae.

pedipalp
The second pair of appendages on the head of spiders, situated between the fangs and the first pair of legs. The first segment of each pedipalp, the coxa, is modified to serve as a chewing mouthpart. The distal segments, collectively called the palp, resemble leg segments in females but in mature male spiders are modified to serve as copulatory organs.

pelagic
adj. Referring to open waters; not in association with the bottom. Epipelagic zone refers to the upper layers of the water column where sunlight penetrates, approximately to a depth of 200 m. Organisms found in deeper waters are often described as mesopelagic (intermediate depths) or bathypelagic (generally >800–1,000 m).

phenotype
Any material aspect of an organism; any measurable or observable organismal trait. Distinguished from the genetic information contained in the genotype of the organism.

pheromone
A chemical signal produced by one member of a species that invokes a response in other members of the same species. Pheromones are used for intraspecific communication.

phylogeny
Evolutionary relationships among groups of organisms indicating hypothesized patterns of descent from common ancestors; the relationships among groups of organisms as determined by their evolutionary history.

phylum
pl. phyla. The taxonomic term for major lineages of animals consisting of all species descended from a single common ancestor. Living animal species are classified into between twenty-eight and thirty-two separate phyla, depending on

the characters used in the classification scheme. I use a widely accepted scheme from the ITIS Catalogue of Life (Bisby et al., 2008) that yields thirty-one separate phyla.

polygyny

An animal mating pattern in which males typically mate with more than one female in a single breeding season.

precocial

adj. Describes the developmental maturity of offspring at birth or hatching. Precocial offspring are capable of independent locomotion and can feed themselves shortly after birth or hatching. Among birds and mammals, precocial offspring have open, functioning eyes and are covered with fur or feathers.

primary sexual trait

A trait that differs between males and females of a species and is part of the reproductive system (gonads plus reproductive tract).

prototroch

The first ring of cilia characteristic of larvae of polychaete annelids.

resource defense polygyny

A mating system or reproductive strategy in which males defend sufficient resources (e.g., food, nest sites, oviposition sites) to attract more than one mate.

secondary sexual trait

A trait that differs between males and females of a species but that is not directly part of the reproductive system.

sessile

adj. Fixed in one position, immobile.

sex

(1) The biological sex of an individual defined on the basis of the type of gamete produced. Males produce smaller, generally more motile gametes (sperm), and females produce larger, more nutrient-rich gametes (eggs). (2) A synonym for genetic recombination or sexual reproduction. (3) In the vernacular can be used as a synonym for sexual intercourse (coitus).

sexual dimorphism

Any consistent morphological difference between males and females within a given species; typically refers to differences in size, shape, color, presence/absence of reproductive organs, weapons for intersexual combat, or sexual display traits; sometimes used more broadly to include any biological differences between sexes including genetic, biochemical, and physiological differences. By convention, a trait that differs on average between males and females is considered to be sexually dimorphic.

sexual reproduction
Reproduction involving ploidy reduction (from diploid to haploid chromosome complements) by means of a cell division process called meiosis and subsequent fusion of two haploid nuclei to form a new diploid nucleus. In animals this process includes production of haploid cells called gametes (eggs and sperm) that unite to form a new diploid individual.

sexual selection
A form of natural selection acting through differential mating success rather than differential survival or fecundity. Mating success in this context means success in fertilizing eggs or in having eggs fertilized.

somatic
adj. From soma, meaning of the body. In biology, refers to bodily traits other than those that comprise the reproductive system (gametes, gonads, and associated organs, ducts, and glands). When referring to cells, somatic cells are distinguished from the germ cells (gametes).

spermatophore
A packet of sperm usually encased in a gelatinous protein capsule that is deposited on the substrate or on the female or is inserted into a specialized genital opening of the female.

symbiotic
adj. Describes a close ecological relationship between the individuals of two different species.

taxa
Plural of taxon. A general term for a named group of related organisms in the hierarchical biological classification system. For animals, typical taxonomic levels include phyla, classes, orders, genera, and species.

trochophore   The larval form characteristic of polychaete annelids possessing three rings of cilia for locomotion.

trophosome   A specialized tissue in marine tubeworms (Annelida, Polychaeta, Siboglinidae) that supports colonies of symbiotic bacteria from which the host worm indirectly obtains its nutrition.

SOURCES

Adiyodi, K. G., and R. G. Adiyodi, eds. 1989. Reproductive Biology of Invertebrates. Volume IV, Part A. Fertilization, Development, and Parental Care. John Wiley & Sons, Chichester, UK.

———, eds. 1990. Reproductive Biology of Vertebrates. Volume IV, Part B. Fertilization, Development and Parental Care. John Wiley & Sons, Chichester, UK.

———, eds. 1992. Reproductive Biology of Invertebrates, Volume V. Sexual Differentiation and Behaviour. John Wiley & Sons, New Delhi, India.

———, eds. 1993. Reproductive Biology of Invertebrates. Volume VI, Part A. Asexual Propagation and Reproductive Strategies. John Wiley & Sons, Chichester, UK.

———, eds. 1994. Reproductive Biology of Invertebrates. Volume VI, Part B. Asexual Propagation and Reproductive Strategies. John Wiley & Sons, Chichester, UK.

Adoutte, A., G. Balavoine, N. Lartillot, O. Lespinet, B. Prud'homme, and R. de Rosa. 2000. The new animal phylogeny: Reliability and implications. Proceedings of the National Academy of Sciences of the United States of America 97:4453–56.

Alexander, R. D., J. L. Hoogland, R. D. Howard, K. M. Noonan, and P. W. Sherman. 1979. Sexual dimorphisms and breeding systems in pinnipeds, ungulates, primates, and humans. Pp. 402–35 in N. A. Chagnon and W. Irons, eds. Evolutionary Biology and Human Social Behavior: An Anthropological Perspective. Duxbury, North Scituate. Massachusetts.

Alonso, J. C., and J. A. Alonso. 1992. Male-biased dispersal in the great bustard. Ornis Scandinavica 23:81–88.

Alonso, J. C., M. Magaña, J. A. Alonso, C. Palacín, C. A. Martín, and B. Martín. 2009. The most extreme sexual size dimorphism among birds: Allometry, selection, and early juvenile development in the great bustard (Otis tarda). Auk 126:657–65.

Alonso, J. C., C. A. Martín, J. A. Alonso, C. Palacín, M. Magaña, and S. J. Lane. 2004. Distribution dynamics of a great bustard metapopulation throughout a decade: Influence of conspecific attraction and recruitment. Biodiversity and Conservation 13:1659–75.

Alonso, J. C., C. A. Martín, J. A. Alonso, C. Palacín, M. Magaña, D. Lieckfeldt, and C. Pitra. 2009. Genetic diversity of the great bustard in Iberia and Morocco: Risks from current population fragmentation. Conservation Genetics 10:379–90.

Alonso, J. C., E. Martín, J. A. Alonso, and M. B. Morales. 1998. Proximate and ultimate causes of natal dispersal in the great bustard *Otis tarda*. Behavioral Ecology 9:245–52.

Alonso, J. C., M. B. Morales, and J. A. Alonso. 2000. Partial migration and lek and nesting area fidelity in female great bustards. The Condor 102:127–36.

Alonso, J. C., C. Palacín, J. A. Alonso, and C. A. Martín. 2009. Post-breeding migration in male great bustards: Low tolerance of the heaviest palearctic bird to summer heat. Behavioral Ecology and Sociobiology 63:1705–15.

Amundsen, T. 2000. Why are female birds ornamented? Trends in Ecology and Evolution 15:149–55.

Amundsen, T., E. Forsgren, and L.T.T. Hansen. 1997. On the function of female ornaments: Male bluethroats prefer colourful females. Proceedings of the Biological Society Washington 264:1579–86.

Anderson, D. T. 1994. Barnacles—Structure, Function, Development and Evolution. Chapman and Hall, London.

Anderson, R. C., J. B. Wood, and R. A. Byrne. 2002. Octopus senescence: The beginning of the end. Journal of Applied Animal Welfare Science 5:275–83.

Andersson, M. 1994. Sexual Selection. Princeton University Press, Princeton, New Jersey.

Andrade, M.C.B. 1996. Sexual selection for male sacrifice in the Australian redback spider. Science 271:70–72.

———. 2003. Risky mate search and male self-sacrifice in redback spiders. Behavioral Ecology 14:531–88.

Antonelis, G. A., M. S. Lowry, C. H. Fiscus, B. S. Stewart, and R. L. DeLong. 1994. Diet of the northern elephant seal. Pp. 211–23 *in* B. J. Le Boeuf and R. M. Laws, eds. Elephant Seals: Population Ecology, Behavior, and Physiology. University of California Press, Berkeley, California.

Arak, A. 1988. Sexual dimorphism in body size: A model and a test. Evolution 42:820–25.

Arnblom, T., M. A. Fedak, and I. L. Boyd. 1997. Factors affecting maternal expenditure in southern elephant seals during lactation. Ecology 78:471–83.

Arnold, J. M., and L. D. Williams-Arnold. 1977. Cephalopoda: Decapoda. Pp. 243–90 *in* A. C. Giese and J. S. Pearse, eds. Reproduction of Marine Invertebrates. Volume IV. Molluscs: Gastropods and Cephalopods. Academic Press, New York.

Arnold, S. J. 1985. Quantitative genetic models of sexual selection. Experientia 41:1296–1309.

Arnold, S. J., and D. Duvall. 1994. Animal mating systems: A synthesis based on selection theory. The American Naturalist 143:317–48.

Arnqvist, G., and S. Henriksson. 1997. Sexual cannibalism in the fishing spider and a model for the evolution of sexual cannibalism based on genetic constraints. Evolutionary Ecology 11:255–73.

Arnqvist, G., and T. Nilsson. 2000. The evolution of polyandry: Multiple mating and female fitness in insects. Animal Behaviour 60:145–64.

Austin, C. R., and R. G. Edwards. 1981. Mechanisms of Sex Differentiation in Animals and Man. Academic Press, New York.

Badyaev, A. V., and G. E. Hill. 2000. The evolution of sexual dimorphism in the house finch. I. Population divergence in morphological covariance structure. Evolution 54:1784–94.

———. 2003. Avian sexual dichromatism in relation to phylogeny and ecology. Annual Review of Ecology, Evolution and Systematics 34:27–49.

Bainbridge, D. 2003. The X in Sex. How the X-Chromosome Controls our Lives. Harvard University Press, Cambridge, Massachusetts.

Baird, P. A., T. W. Andersen, H. B. Newcombe, and R. B. Lowry. 1988. Genetic disorders in children and young adults: A population study. American Journal of Human Genetics 42:677–93.

Baird, R. C., and G. Y. Jumper. 1995. Encounter models and deep-sea fishes: Numerical simulations and the mate location problem in *Sternoptyx diaphana* (Pisces, Sternoptychidae). Deep Sea Research Part I: Oceanographic Research Papers 42:675–96.

Barlow, G. W. 2000. The Cichlid Fishes. Perseus Publishing, Cambridge, Massachusetts.

Baron-Cohen, S. 2004. The Essential Difference: Men, Women and the Extreme Male Brain. Penguin Books, New York.

Barth, F. G. 2002. A Spider's World. Senses and Behavior. Springer-Verlag, Berlin.

Bayly, I.A.E. 1978. Variation in sexual size dimorphism in nonmarine calanoid copepods and its ecological significance. Limnology and Oceanography 23:1224–28.

Bell, G. 1982. The Masterpiece of Nature. University of California Press, Berkeley and Los Angeles, California.

Bel-Venner, M. C., S. Dray, D. Allainé, F. Menu, and S. Venner. 2008. Unexpected male choosiness for mates in a spider. Proceedings of the Royal Society B. 275:77–82.

Bennett, A.T.D., I. C. Cuthill, and K. J. Norris. 1994. Sexual selection and the mismeasure of color. American Naturalist 144:848–60.

Benz, G. W., and G. B. Deets. 1986. *Kroyeria caseyi* sp. nov. (Kroyeriidae: Siphonostomatoida), a parasitic copepod infesting gills of night sharks (*Carcharhinus signatus* [Poey, 1868]) in the western North Atlantic. Canadian Journal of Zoology 64:2492–98.

Berec, L., P. J. Schembri, and D. S. Boukal. 2005. Sex determination in *Bonellia viridis* (Echiura: Bonelliidae): population dynamics and evolution. Oikos 108:473–84.

Berglund, A., C. Magnhagen, A. Bisazza, B. Koenig, and F. Huntingford. 1993. Female-female competition over reproduction. Behavioral Ecology 4:184–87.

Berglund, A., and G. Rosenqvist. 2001. Male pipefish prefer dominant over attractive females. Behavioral Ecology 12:402–6.

Bertelsen, E., ed. 1951. The Ceratioid Anglerfishes. Ontogeny, Distribution and Biology. Bianco Luno, Copenhagen.

Bininda-Edmonds, O.R.P., and J. L. Gittleman. 2000. Are pinnipeds functionally different from fissiped carnivores? The importance of phylogenetic comparative analyses. Evolution 54:1011–23.

Bird, A. F., and R. I. Sommerville. 1989. Nematoda and Nematomorpha. Pp. 219–50 *in* K. G. Adiyodi and R. G. Adiyodi, eds. Reproductive Biology of Invertebrates. Volume IV. Part A. Fertilization, Development and Parental Care. John Wiley & Sons, Chichester, UK.

Birkhead, T. R., and A. P. Moller. 1998. Sperm Competition and Sexual Selection. Academic Press, San Diego, California.

Bisby, F. A., Y. R. Roskov, T. M. Orrell, D. Nicolson, L. E. Paglinawan, N. Bailly, P. M. Kirk, T. Bourgoin, and J. van Hertum. 2008. Species 2000 & ITIS Catalogue of Life: 2008 Annual Checklist. Digital resource at www.catalogueoflife.org/annual-checklist/2008/. Species 2000: Reading, UK.

Blackwelder, R. E., and B. A. Shepherd. 1981. The Diversity of Animal Reproduction. CRC Press, Boca Raton, Florida.

Blanckenhorn, W. 2000. The evolution of body size: What keeps organisms small? Quarterly Review of Biology 75:385–407.

———. 2005. Behavioral causes and consequences of sexual size dimorphism. Ethology 111:977–1016.

Blanckenhorn, W. U., A. F. Dixon, D. J. Fairbairn, M. W. Foellmer, P. Gibert, K. van der Linde, R. Meier, S. Nylin, S. Pitnick, C. Schoff, M. Signorelli, T. Teder, and C. Wiklund. 2007. Proximate causes of Rensch's rule: Does sexual size dimorphism in arthropods result from sex differences in development time? American Naturalist 169:245–57.

Blanckenhorn, W. U., R. Meier, and T. Teder. 2007. Rensch's rule in insects: Patterns among and within species. Pp. 60–70 *in* D. J. Fairbairn, W. U. Blanckenhorn, and T. Székely, eds. Sex, Size and Gender Roles. Evolutionary Studies of Sexual Size Dimorphism. Oxford University Press, Oxford, UK.

Blum, D. 1997. Sex on the Brain. The Biological Differences between Men and Women. Viking, Penguin Putman, New York.

Blüm, V. 1985. Vertebrate Reproduction. A Textbook. Springer-Verlag, Berlin.

Boggs, C. L., W. B. Watt, and P. R. Ehrlich. 2003. Butterflies. Ecology and Evolution Taking Flight. University of Chicago Press, Chicago, London.

Boletzky, S. v. 2003. A lower limit to adult size in coleoid cephalopods: Elements of a discussion. Berliner Palaobiologische Abhandlungen 3:19–28.

Bonduriansky, R. 2001. The evolution of male mate choice in insects: A synthesis of ideas and evidence. Biological Reviews 76:305–39.

Bonner, N. 1994. Seals and Sea Lions of the World. Facts on File, New York.

Boyd, I. L., T. A. Arnbom, and M. A. Fedak. 1994. Biomass and energy consumption of the South Georgia population of southern elephant seals. Pp. 98–120 *in* B. J. Le Boeuf and R. M. Laws, eds. Elephant Seals: Population Ecology, Behavior, and Physiology. University of California Press, Berkeley, California.

Boyle, P. R. 1987. Cephalopod Life Cycles. Volume II. Comparative Reviews. Academic Press, London.

Braby, C. E., G. W. Rouse, S. B. Johnson, W. J. Jones, and R. C. Vrijenhoek. 2007. Bathymetric and temporal variation among *Osedax* boneworms and associated megafauna on whale-falls in Monterey Bay, California. Deep-Sea Research Part I—Oceanographic Research Papers 54:1773–91.

Bradbury, J. W., and M. B. Andersson. 1987. Sexual Selection: Testing the Alternatives. John Wiley & Sons, New York.

Brana, F. 1996. Sexual dimorphism in lacertid lizards: Male head increase vs. female abdomen increase? Oikos 75:511–23.

Breder, C. M., and D. E. Rosen. 1966. Modes of Reproduction in Fishes. The American Museum of Natural History, Garden City, New York.

Brizendine, L. 2006. The Female Brain. Morgan Road Books, New York.

Brooks, R., and J. A. Endler. 2001. Direct and indirect sexual selection and quantitative genetics of male traits in guppies (*Poecilia reticulata*). Evolution 55:1002–15.

Brown, J. H., and G. B. West. 2000. Scaling in Biology. Oxford University Press, New York.

Bull, J. J. 1980. Sex determination in reptiles. The Quarterly Review of Biology 55:3–21.

———. 2008. Sex determination: Are two mechanisms better than one? Journal of Bioscience 32:5–8.

Calder, W.A.I. 1984. Size, Function, and Life History. Harvard University Press, Cambridge, Massachusetts.

Carlini, A. R., S. Poljak, G. A. Daneri, M.E.I. Márquez, and J. Negrete. 2006. The dynamics of male harem dominance in southern elephant seals (*Mirounga leonina*) at the South Shetland Islands. Polar Biology 29:796–805.

Carranza, J., and S. J. Hidalgo de Trucios. 1993. Condition-dependence and sex traits in the male great bustard. Ethology 94:187–200.

Carranza, J., S. J. Hidalgo de Trucios, and V. Ena. 1989. Mating system flexibility in the great bustard: A comparative study. Bird Study 36:192–98.

Carrier, J. C., J. A. Musick, and M. R. Heithans. 2004. Biology of Sharks and Their Relatives. CRC Press, Boca Raton, Florida.

Carroll, R. L. 1987. Vertebrate Paleontology and Evolution. Freeman & Co., New York.

Chae, J., and S. Nishida. 1994. Integumental ultrastructure and color patterns in the iridescent copepods of the family Sapphirinidae (Copepoda: Poecilostomatoida). Marine Biology 119:205–10.

Chanley, P., and M. H. Chanley. 1970. Larval development of the commensal clam, *Montacuta percompressa* Dall. Proceedings of the Malacological Society London 39:59–67.

Charnov, E. 1982. The Theory of Sex Allocation. Princeton University Press, Princeton, New Jersey.

Cherel, Y., S. Ducatez, C. Fontain, P. Richard, and C. Guinet. 2008. Stable isotopes reveal the trophic position and mesopelagic fish diet of female southern elephant seals breeding on the Kerguelen Islands. Marine Ecology Progress Series 370:239–47.

Christenson, T. E., and K. C. Goist. 1979. Costs and benefits of male–male competition in the orb weaving spider, *Nephila clavipes*. Behavioral Ecology and Sociobiology 5:87–92.

Chronin, H. 1991. The Ant and the Peacock. Cambridge University Press, Cambridge, New York, Melbourne.

Clarke, T. A. 1983. Sex ratios and sexual differences in size among mesopelagic fishes from the Central Pacific Ocean. Marine Biology 73:203–9.

Claxton, S. 1996. Sexual dimorphism in Australian *Echiniscus* (Tardigrada, Echiniscidae) with descriptions of three new species. Zoological Journal of the Linnean Society 116:13–33.

Clinton, W. L. 1994. Sexual selection and growth in male northern elephant seals. Pp. 154–68 *in* B. J. Le Boeuf and R. M. Laws, eds. Elephant Seals: Population Ecology, Behavior, and Physiology. University of California Press, Berkeley, California.

Clutton-Brock, T. H., ed. 1988. Reproductive Success. Studies of Individual Variation in Contrasting Breeding Systems. University of Chicago Press, Chicago.

———. 1991. The Evolution of Parental Care. Princeton University Press, Princeton, New Jersey.

———. 2009. Sexual selection in females. Animal Behaviour 77:2–11.

Clutton-Brock, T. H., S. D. Albon, and F. E. Guinness. 1988. Reproductive success in male and female red deer. Pp. 325–43 *in* T. H. Clutton-Brock, ed. Reproductive Success. Studies of Individual Variation in Contrasting Breeding Systems. University of Chicago Press, Chicago.

Clutton-Brock, T. H., J. C. Deutsch, and T.J.C. Nefdt. 1993. The evolution of ungulate leks. Animal Behaviour 46:1121–38.

Cochran, P. A., A. K. Newton, and C. Korte. 2004. Great Gordian knots: Sex ratio and sexual size dimorphism in aggregations of horsehair worms (*Gordius difficilis*). Invertebrate Biology 123:78–82.

Coddington, J. A., G. Hormiga, and N. Scharff. 1997. Giant female or dwarf male spiders? Nature 385:687–88.

Coe, W. R. 1944. .Sexual Differentiation in Mollusks. II. Gastropods, Amphineurans, Scaphopods, and Cephalopods. Quarterly Review of Biology 19:85–97.

Colwell, R. K. 2000. Rensch's rule crosses the line: Convergent allometry of sexual size dimorphism in hummingbirds and flower mites. American Naturalist 156:495–510.

Conn, D. B. 2000. Atlas of Invertebrate Reproduction and Development. Wiley-Liss, New York.

Cooper, V. J., and G. R. Hosey. 2003. Sexual dichromatism and female preference in *Eulemur fulvus* subspecies. International Journal of Primatology 24:1177–88.

Corcobado, G., M. A. Rodríguez-Gironéz, E. De Mas, and J. Moya-Laraño. 2010. Introducing the refined gravity hypothesis of extreme sexual size dimorphism. BMC Evolutionary Biology 10:236.

Cortez, T., B. G. Castro, and A. Guerra. 1995. Reproduction and condition of female *Octopus mimus* (Mollusca: Cephalopoda). Marine Biology 123:505–10.

Costanzo, K., and A. Monteiro. 2007. The use of chemical and visual cues in female choice in the butterfly *Bicyclus anynana*. Proceedings of the Royal Society B 274:845–51.

Cotton, S., K. Fowler, and A. Pomiankowski. 2004. Do sexual ornaments demonstrate heightened condition-dependent expression as predicted by the handicap hypothesis? Proceedings of the Royal Society B 271: 771–83.

Courtiol, A., M. Raymond, B. Godelle, and J.-B. Ferdy. 2010. Mate choice and human stature: Homogamy as a unified framework for understanding mating preferences. Evolution 64:2189–2203.

Cox, R. M., M. A. Butler, and H. B. John-Alder. 2007. The evolution of sexual size dimorphism in reptiles. Pp. 38–49 *in* D. J. Fairbairn, W. U. Blanckenhorn, and T. Székely, eds. Sex, Size and Gender Roles. Evolutionary Studies of Sexual Size Dimorphism. Oxford University Press, Oxford, UK.

Cox, R. M., and R. Calsbeek. 2009. Sexually antagonistic selection, sexual dimorphism, and the resolution of intralocus sexual conflict. American Naturalist 173:176–87.

Crocker, D. E., J. D. Williams, D. P. Costa, and B. J. Le Boeuf. 2001. Maternal traits and reproductive effort in northern elephant seals. Ecology 82:3541–55.

Croft, D. P., M. S. Botham, and J. Krausse. 2004. Is sexual segregation in the guppy, *Poecilia reticulata*, consistent with the predation risk hypothesis? Environmental Biology of Fishes 71:127–33.

Crozier, W. J. 1920. Sex-correlated coloration in *Chiton tuberculatus*. American Naturalist 54:84–88.

Crozier, W. W. 1989. Age and growth of angler-fish (*Lophius piscatorius* L.) in the North Irish Sea. Fisheries Research 7:267–78.

Dallai, R., P. P. Fanciulli, and F. Frati. 1999. Chromosome elimination and sex determination in springtails (Insecta, Collembola). Journal of Experimental Zoology Part B: Molecular and Developmental Evolution 285:215–25.

Darwin, C. 1851. A Monograph of the Sub-class Cirripedia, with Figures of All Species. The Lepadidae; or, Pedunculated Cirripedes. 1. The Ray Society, London.

———. 1854. A Monograph of the Sub-class Cirripedia, with Figures of All the Species. The Balanidae, (or Sessile Cirripedes); the Verrucidae. 2. The Ray Society, London.

———. 1859. The Origin of Species. John Murray, London.

———. 1871. The Descent of Man and Selection in Relation to Sex, Volumes 1 and 2. John Murray, London.

Degnan, S. M., and B. M. Degnan. 2006. The origin of the pelagobenthic metazoan life cycle: What's sex got to do with it? Integrative and Comparative Biology 46:683–90.

De Mas, E., C. Ribera, and J. Moya-Laraño. 2009. Resurrecting the differential mortality model of sexual size dimorphism. Journal of Evolutionary Biology 22:1739–49.

de Meeûs, T., F. Prugnolle, and P. Agnew. 2007. Asexual reproduction: Genetics and evolutionary aspects. Cellular and Molecular Life Sciences 64:1355–72.

Detto, T., P.R.Y. Backwell, J. M. Hemmi, and J. Zeil. 2006. Visually mediated species and neighbour recognition in fiddler crabs (*Uca mjoebergi* and *Uca capricornis*). Proceedings of the Royal Society B 273:1661–66.

Deutsch, C., B. J. Le Boeuf, and D. P. Costa. 1994. Sex differences in reproductive effort in northern elephant seals. Pp. 169–210 *in* B. J. Le Boeuf and R. M. Laws, eds. Elephant Seals: Population Ecology, Behavior, and Physiology. University of California Press, Berkeley, California.

Deutsch, C. J., M. P. Haley, and B. J. Le Boeuf. 1990. Reproductive effort of male northern elephant seals: Estimates from mass loss. Canadian Journal of Zoology 68:2580–93.

Días, E. R., and M. Thiel. 2004. Chemical and visual communication during mate searching in rock shrimp. Biological Bulletin 206:134–43.

Döhren, J., and T. Bartolomaeus. 2006. Ultrastructure of sperm and male reproductive system in *Lineus viridis* (Heteronemertea, Nemertea). Zoomorphology 125:175–85.

Dunn, P. O., L. A. Whittingham, and T. E. Pitcher. 2001. Mating systems, sperm competition, and the evolution of sexual dimorphism in birds. Evolution 55:161–75.

Eberhard, W. G. 1996. Female Control: Sexual Selection by Cryptic Female Choice. Princeton University Press, Princeton, New Jersey.

———. 2009. Postcopulatory sexual selection: Darwin's omission and its consequences. Proceedings of the National Academy of Sciences of the United States of America 106:10025–32.

Eberhard, W. G., and C. Cordero. 1995. Sexual selection by cryptic female choice on male seminal products—a new bridge between sexual selection and reproductive physiology. Trends in Ecology and Evolution 10:493–96.

Eberhard, W. G., and B. A. Huber. 2010. Spider genitalia. Precise maneuvers with a numb structure in a complex lock. Pp. 249–84 *in* J. L. Leonard and A. Córdoba-Aquilar, eds. The Evolution of Primary Sexual Characters in Animals. Oxford University Press, New York.

Eggert, C. 2004. Sex determination: the amphibian models. Reproduction Nutrition Development 44:539–49.

Elder, H. Y. 1979. Studies on the host parasite relationship between the parasitic prosobranch *Thyca crystallina* and the asteroid starfish *Linckia laevigata*. Journal of Zoology London 187:369–91.

Elgar, M. A. 1991. Sexual cannibalism, size dimorphism, and courtship behavior in orb-weaving spiders (Aranidae). Evolution 45:444–48.

Elgar, M. A., and B. F. Fahey. 1996. Sexual cannibalism, competition, and size dimorphism in the orb-weaving spider *Nephila plumipes* Latreille (Araneae: Araneoidea). Behavioral Ecology 7:195–98.

Elgar, M. A., and D. R. Nash. 1988. Sexual cannibalism in the garden spider *Araneus diadematus*. Animal Behaviour 36:1511–17.

Elgar, M. A., J. M. Schneider, and M. E. Herberstein. 2000. Female control of paternity in the sexually cannibalistic spider *Argiope keyserlingi*. Proceedings of the Royal Society B 267:2439–43.

Eliot, L. 2010. The truth about boys and girls. Scientific American Mind, May/June 2010, pp. 22–29.

Emlen, D. J. 2008. The evolution of animal weapons. Annual Review of Ecology Evolution and Systematics 39:387–413.

Emlen, D. J., and F. Nijhout. 2000. The development and evolution of exaggerated morphologies in insects. Annual Review of Entomology 45:661–708.

Emlet, R. B. 2002. Ecology of adult sea urchins. Pp. 111–15 *in* Y. Yokota, V. Matranga, and Z. Smolenicka, eds. The Sea Urchin: From Basic Biology to Aquaculture. A. A. Balkema, Swets and Zeitlinger, Lisse, The Netherlands.

Ena, V., A. Martinez, and D. H. Thomas. 1987. Breeding success of the Great Bustard *Otis tarda* in Zamora Province, Spain, in 1984. Ibis 129:364–70.

Enders, F. 1977. Web-site selection by orb-web spiders, particularly *Argiope aurantia* Lucas. Animal Behaviour 25:694–712.

Endler, J. A. 1992. Signal, signal conditions, and the direction of evolution. The American Naturalist 139:S125–53.

———. 1993. The color of light in forests and its implications. Ecological Monographs 63:1–27.

Endler, J. A., and A. L. Basolo. 1998. Sensory ecology, receiver bias and sexual selection. Trends in Ecology and Evolution 13:415–20.

Endler, J. A., and A. Houde. 1995. Geographic variation in female preferences for male traits in *Poecilia reticulata*. Evolution 49:456–68.

Endler, J. A., and P. W. Mielke. 2005. Comparing entire colour patterns as birds see them. Biological Journal of the Linnean Society 86:405–31.

Epp, R. W., and W.M.J. Lewis. 1979. Sexual dimorphism in *Brachionus plicatilis* (Rotifera): Evolutionary and adaptive significance. Evolution 33:919–28.

Erlandsson, A., and A. J. Ribbink. 1997. Patterns of sexual size dimorphism in African cichlid fishes. South African Journal of Science 93:498–508.

Esperk, T., T. Tammaru, S. Nylin, and T. Teder. 2007. Achieving high sexual size dimorphism in insects: Females add instars. Ecological Entomology 32:243–56.

Fabiani, A., F. Galimberti, S. Sanvito, and A. R. Hoelzel. 2004. Extreme polygyny among southern elephant seals on Sea Lion Island, Falkland Islands. Behavioral Ecology 15:961–69.

———. 2006. Relatedness and site fidelity at the southern elephant seal, *Mirounga leonina*, breeding colony in the Falkland Islands. Animal Behaviour 72:617–26.

Fairbairn, D. J. 1977a. The spring decline in deermice: Death or dispersal? Canadian Journal of Zoology 55:84–92.

———. 1977b. Why breed early? A study of reproductive tactics in *Peromyscus maniculatus*. Canadian Journal of Zoology 55:862–71.

———. 1978a. Behaviour of dispersing deer mice (*Peromyscus maniculatus*). Behavioral Ecology and Sociobiology 3:265–82.

Fairbairn, D. J. 1978b. Dispersal of deermice, *Peromyscus maniculatus*: Proximal causes and effects on fitness. Oecologia 32:171–93.

———. 1988. Sexual selection for homogamy in the Gerridae: An extension of Ridley's comparative approach. Evolution 42:1212–22.

———. 1997. Allometry for sexual size dimorphism: Pattern and process in the coevolution of body size in males and females. Annual Review of Ecology and Systematics 28:659–87.

———. 2013. Data from: Odd Couples. Extraordinary Differences between the Sexes in the Animal Kingdom. Archived in the Dryad repository: http://dx.dio.org/10.5061/dryad.n48cm.

Fairbairn, D. J., W. U. Blanckenhorn, and T. Székely, eds. 2007. Sex, Size and Gender Roles. Evolutionary Studies of Sexual Size Dimorphism. Oxford University Press, Oxford, UK.

Fairbairn, D. J., and D. A. Roff. 2006. The quantitative genetics of sexual dimorphism: Assessing the importance of sex-linkage. Heredity 97:319–28.

Fausto-Sterling, A., P. Gowaty, and M. Zuk. 1997. Evolutionary psychology and Darwinian feminism. Feminist Studies 23:403–17.

Fedak, M. A., T. A. Arnbom, B. J. McConnell, C. Chambers, I. L. Boyd, J. Harwood, and T. S. McCann. 1994. Expenditure, investment, and acquisition of energy in southern elephant seals. Pp. 354–73 *in* B. J. Le Boeuf and R. M. Laws, eds. Elephant Seals: Population Ecology, Behavior, and Physiology. University of California Press, Berkeley, California.

Felsenstein, J. 1974. The evolutionary advantage of recombination. Genetics 78:737–56.

Ferguson-Smith, M. 2007. The evolution of sex chromosomes and sex determination in vertebrates and the key role of DMRT1. Sexual Development 1:2–11.

Festa-Bianchet, M. 1991. The social system of bighorn sheep: Grouping patterns, kinship and female dominance rank. Animal Behaviour 42:71–82.

ffrench-Constant, R., and P. B. Koch. 2003. Mimicry and melanism in swallowtail butterflies: Toward a molecular understanding. Pp. 281–319 *in* C. L. Boggs, W. B. Watt, and P. R. Ehrlich, eds. Butterflies. Ecology and Evolution Taking Flight. University of Chicago Press, Chicago, London.

Field, I. C., C.J.A. Bradshaw, H. R. Burton, and M. A. Hindell. 2007. Differential resource allocation strategies in juvenile elephant seals in the highly seasonal Southern Ocean. Marine Ecology Progress Series 331:281–90.

Filiz, H., and E. Taskavak. 2006. Sexual dimorphism in the head, mouth, and body morphology of the small-spotted catshark, *Scyliorhinus canicula* (Linnaeus, 1758) (Chondrichthyes: Scyliohinidae) from Turkey. Acta Adriatica 47:37–47.

Fine, C. 2010. Delusions of Gender. How Our Minds, Society, and Neurosexism Create Difference. W. W. Norton, New York.

Fiske, P., P. T. Rintamaki, and E. Karvonen. 1998. Mating success in lekking males: A meta-analysis. Behavioral Ecology 9:328–38.

Foelix, R. F. 1996. Biology of Spiders, 2nd edition. Oxford University Press, New York.

Foellmer, M. W. 2004. Sexual dimorphism and sexual selection in the highly dimorphic orb-weaving spider *Argiope aurantia* (Lucas). PhD Thesis. Concordia University, Montreal, Quebec, Canada.

———. 2008. Broken genitals function as mating plugs and affect sex ratios in the orb-web spider *Argiope aurantia*. Evolutionary Ecology Research 10:449–62.

Foellmer, M. W., and D. J. Fairbairn. 2003. Spontaneous male death during copulation in an orb-weaving spider. Proceedings of the Royal Society London B (Supplement) 270:S183–85.

———. 2004. Males under attack: Sexual cannibalism and its consequences for male morphology and behavior in an orb-weaving spider. Evolutionary Ecology Research 6:163–81.

———. 2005a. Competing dwarf males: Sexual selection in an orb-weaving spider. Journal of Evolutionary Biology 18:629–41.

———. 2005b. Selection on male size, leg length and condition during mate search in a sexually highly dimorphic orb-weaving spider. Oecologia 142:653–62.

Foellmer, M. W., and Moya-Laraño, J. 2007. Sexual size dimorphism in spiders: Patterns and processes. Pp. 71–81 *in* D. J. Fairbairn, W. U. Blanckenhorn, and T. Székely, eds. Sex, Size and Gender Roles. Evolutionary Studies of Sexual Size Dimorphism. Oxford University Press, Oxford, UK.

Fromhage, L., M. A. Elgar, and J. M. Schneider. 2005. Faithful without care: The evolution of monogyny. Evolution 59:1400–5.

Fromhage, L., J. M. McNamara, and A. I. Houston. 2008. A model for the evolutionary maintenance of monogyny in spiders. Journal of Theoretical Biology 250:524–31.

Fudge, S. B., and G. A. Rose. 2008. Life history co-variation in a fishery depleted Atlantic cod stock. Fisheries Research 92:107–13.

Fujikura, K., Y. Fujiwara, and M. Kawato. 2006. A new species of *Osedax* (Annelida: Siboglinidae) associated with whale carcasses off Kyushu, Japan. Zoological Science 23:733–40.

Galimberti, F., L. Boitani, and I. Marzetti. 2000a. Female strategies of harassment reduction in southern elephant seals. Ethology, Ecology and Evolution 12:367–88.

———. 2000b. Harassment during arrival on land and departure to sea in southern elephant seals. Ethology, Ecology and Evolution 12:389–404.

———. 2000c. The frequency and costs of harassment in southern elephant seals. Ethology, Ecology and Evolution 12:345–65.

Galimberti, F., A. Fabiani, and S. Sanvito. 2002a. Measures of breeding inequality: A case study in southern elephant seals. Canadian Journal of Zoology 80:1240–49.

———. 2002b. Opportunity for selection in southern elephant seals (*Mirounga leonina*): The effect of spatial scale of analysis. Journal of Zoology 256:93–97.

Galimberti, F., S. Sanvito, C. Braschi, and L. Boitani. 2007. The cost of success: Reproductive effort in male southern elephant seals (*Mirounga leonina*). Behavioral Ecology and Sociobiology 62:159–71.

Garey, J. R., S. J. McInnes, and B. Nichols. 2008. Global diversity of tardigrades (Tardigrada) in freshwater. Hydrobiologia 595:101–6.

Garm, A., M. M. Coates, R. Gad, J. Seymour, and D. E. Nilsson. 2007. The lens eyes of the box jellyfish *Tripedalia cystophora* and *Chiropsalmus* sp. are slow and color-blind. Journal of Comparative Physiology A, Neuroethology, Sensory, Neural and Behavioral Physiology 193:547–57.

Gaskett, A. C. 2007. Spider sex pheromones: Emission, reception, structures, and functions. Biological Reviews 82:26–48.

Geary, D. C. 1998. Male, Female. The Evolution of Human Sex Differences. American Psychological Association, Washington, D.C.

Geddes, M. C., and G. A. Cole. 1981. Variation in sexual size differentiation in North American diaptomids (Copepoda: Calanoida): Does variation in the degree of dimorphism have ecological significance? Limnology and Oceanography 26:367–74.

Geist, V. 1971. Mountain Sheep. University of Chicago Press, Chicago.

Gertsch, W. J. 1979. American Spiders. Van Nostrand Reinhold, New York.

Ghiselin, M. T. 1974. The Economy of Nature and the Evolution of Sex. University of California Press, Berkeley, California.

Gibbons, W. J., and J. E. Lovich. 1990. Sexual dimorphism in turtles with emphasis on the slider turtle (*Trachemys scripta*). Herpetological Monographs 4:1–29.

Giese, A. C., and J. S. Pearse, eds. 1974. Reproduction of Marine Invertebrates. Volume I. Aocoelomate and Pseudocoelomate Metazoans. Academic Press, New York.

———, eds. 1975a. Reproduction of Marine Invertebrates. Volume II. Entoprocts and Lesser Coelomates. Academic Press, New York.

———, eds. 1975b. Reproduction of Marine Invertebrates, Volume III Annelids and Echiurans. Academic Press, New York.

———, eds. 1977. Reproduction of Marine Invertebrates. Volume IV. Molluscs: Gastropods and Cephalopods. Academic Press, New York.

———, eds. 1979. Reproduction of Marine Invertebrates. Volume V. Molluscs: Pelecypods and Lesser Classes. Academic Press, New York.

Gilbert, J. J. 1989. Rotifera. Pp. 179–99 *in* K. G. Adiyodi and R. G. Adiyodi, eds. Reproductive Biology of Invertebrates. Volume IV. Part A. Fertilization, Development, and Parental Care. John Wiley & Sons, Chichester, UK.

Gilbert, J. J., and C. E. Williamson. 1983. Sexual dimorphism in zooplankton (Copepods, Cladocera, and Rotifera). Annual Review of Ecology and Systematics 14:1–33.

Gissler, C. F. 1882. *Boptrus manhattensis* from the gill cavity of *Palemonetes vulgaris* Stimpson. Proceedings of the American Association for the Advancement of Science, 30th meeting, pp. 243–45.

Glover, A. G., B. Kallstrom, C. R. Smith, and T. G. Dahlgren. 2005. Worldwide whale worms? A new species of *Osedax* from the shallow North Atlantic. Proceedings of the Royal Society B 272:2587–92.

Goffredi, S. K., S. B. Johnson, and R. C. Vrijenhoek. 2007. Genetic diversity and potential function of microbial symbionts associated with newly discovered species of *Osedax* polychaete worms. Applied and Environmental Microbiology 73:2314–23.

Goffredi, S. K., V. J. Orphan, G. W. Rouse, L. Jahnke, T. Embaye, K. Turk, R. Lee, and R. C. Vrijenhoek. 2005. Evolutionary innovation: A bone-eating marine symbiosis. Environmental Microbiology 7:1369–78.

Goffredi, S. K., C. K. Paull, K. Fulton-Bennett, L. A. Hurtado, and R. C. Vrijen-
hoek. 2004. Unusual benthic fauna associated with a whale fall in Monterey
Canyon, California. Deep-Sea Research Part I—Oceanographic Research
Papers 51:1295–1306.

Gomez, D., and M. Théry. 2007. Simultaneous crypsis and conspicuousness in
color patterns: Comparative analysis of a neotropical rainforest bird community. American Naturalist 169:542–61.

Gonor, J. J. 1979. Monoplacophora. Pp. 87–93 in A. C. Giese and J. S. Pearse,
eds. Reproduction of Marine Invertebrates. Volume V. Molluscs: Pelecypods
and Lesser Classes. Academic Press, New York.

Gonzalez-Voyer, A., J. L. Fitzpatrick, and N. Kolm. 2008. Sexual selection determines parental care patterns in cichlid fishes. Evolution 62:2015–26.

Gotelli, N. J., and H. R. Spivey. 1992. Male parasitism and intrasexual competition in a burrowing barnacle. Oecologia 91:474–80.

Gray, D. A. 1996. Carotenoids and sexual dichromatism in North American
passerine birds. The American Naturalist 148:453–80.

Greenwood, P. J., and J. Adams. 1987. The Ecology of Sex. Edward Arnold, London.

Guiso, L., M. Ferdinando, P. Sapienza, and L. Zingales. 2008. Culture, gender,
and math. Science 320:1164–65.

Guldberg, H. J., and R. M. Kristensen. 2006. The "hyena female" of tardigrades
and descriptions of two new species of Megastygarctides (Arthrotardigrada:
Stygarctidae) from Saudi Arabia. Hydrobiologia 558:81–101.

Gustafsson, A., and P. Lindenfors. 2004. Human size evolution: No evolutionary allometric relationship between male and female stature. Journal of
Human Evolution 47:253–66.

Haag, E. S., and A. V. Doty. 2005. Sex determination across evolution: Connecting the dots. PLoS Biology 3(1): e21. doi:10.1371/journal.pbio.0030021

Hadfield, M. G. 1975. Hemichordata. Pp. 185–240 in A. C. Giese and J. S.
Pearse, eds. Reproduction of Marine Invertebrates. Vol. II. Entoprocts and
Lesser Coelomates. Academic Press, New York.

———. 1979. Aplacophora. Pp. 1–26 in A. C. Giese and J. S. Pearse, eds. Reproduction of Marine Invertebrates. Volume V. Molluscs: Pelecypods and Lesser
Classes. Academic Press, New York.

Haley, M. P., C. J. Deutsch, and B. J. Le Boeuf. 1994. Size, dominance and copulatory success in male northern elephant seals, *Mirounga angustirostris*. Animal Behaviour 48:1249–60.

Halliday, T. 1980. Sexual Strategy. University of Chicago Press, Chicago.

Hamel, J.-F., and J. H. Himmelman. 1992. Sexual dimorphism in the sand dollar *Echinarachnius parma*. Marine Biology 113:379–83.

Hamilton, W. D. 1990. Mate choice near and far. American Zoologist 30: 341–52.

Hammond, G. 2002. "*Argiope aurantia*" (Online), Animal Diversity Web, http://animaldiversity.ummz.umich.edu/site/accounts/information/ Argiope_aurantia.html. Accessed November 10, 2010.

Hanlon, R. T., and J. B. Messenger. 1996. Cephalopod Behaviour. Cambridge University Press, Cambridge, UK.

Harbison, G. R., and R. L. Miller. 1986. Not all ctenophores are hermaphrodites. Studies on the systematics, distribution, sexuality and development of two species of *Ocyropsis*. Marine Biology 90:413–24.

Harwood, R. H. 1974. Predatory behavior of *Argiope aurantia* (Lucas). American Midland Naturalist 91:130–39.

Hayes, T. B. 1998. Sex determination and primary sex differentiation in amphibians: Genetic and developmental mechanisms. Journal of Experimental Zoology, Part A: Comparative Experimental Zoology 281:373–99.

Head, G. 1995. Selection on fecundity and variation in the degree of sexual size dimorphism among spider species (class Aranae). Evolution 49:776–81.

Hebert, P.D.N. 1977. A revision of the taxonomy of the genus *Daphnia* (Crustacea: Daphnidae) in south-eastern Australia. Australian Journal of Zoology 25:371–98.

Hedrick, A. V., and E. J. Temeles. 1989. The evolution of sexual dimorphism in animals: Hypotheses and tests. Trends in Ecology and Evolution 4:136–38.

Hegyi, G., B. Rosivali, E. Szöllösi, R. Hargita, M. Eens, and J. Török. 2007. A role for female ornamentation in the facultatively polygynous mating system of collared flycatchers. Behavioral Ecology 18:1116–1122.

Heinroth, O., and M. Heinroth. 1927. Die Vögel Mitteleuropas, Volume 3. Lichterfelda, Berlin.

Heller, J. 1993. Hermaphroditism in molluscs. Biological Journal of the Linnean Society 48:19–42.

Herberstein, M. E., J. M. Schneider, A.M.T. Harmer, A. C. Gaskett, K. Robinson, K. Shaddick, D. Soetkamp, P. D. Wilson, S. Pekar, and M. A. Elgar. 2011. Sperm storage and copulation duration in a sexually cannibalistic spider. Journal of Ethology 29:9–15.

Hickman, C. P., L. S. Roberts, S. L. Keen, A. Larson, and D. J. Eisenhour. 2007. Animal Diversity, 4th edition. McGraw-Hill, New York.

Hidalgo de Trucios, S. J., and J. Carranza. 1991. Timing, structure and functions of the courtship display in male great bustard. Ornis Scandinavica 22:360–66.

Hilario, A., M. Capa, T. G. Dahlgren, K. M. Halanych, C.T.S. Little, D. J. Thornhill, C. Verna, and A. G. Glover. 2011. New perspectives on the ecology and evolution of siboglinid tubeworms. PLoS One e16309. doi: 10.1371/journal.pone.0016309.

Hindell, M. A., D. J. Slip, and H. R. Burtin. 1991. The diving behavior of adult male and female southern elephant seals, *Mirounga leonina* (Pinnipedia, Phocidae). Australian Journal of Zoology 39:595–619.

Hogg, J. T. 1987. Intrasexual competition and mate choice in Rocky Mountain bighorn sheep. Ethology 75:119–44.

Hogg, J. T., and S. H. Forbes. 1997. Mating in bighorn sheep: frequent male reproduction via a high-risk "unconventional" tactic. Behavioral Ecology and Sociobiology 41:33–48.

Höglund, J., and R. V. Alatalo. 1995. Leks. Princeton University Press, Princeton, New Jersey.

Höglund, J., and B. Sillén-Tullberg. 1994. Does lekking promote the evolution of male-biased size dimorphism in birds? On the use of comparative approaches. The American Naturalist 144:881–89.

Holden, C., and R. Mace. 1999. Sexual dimorphism in stature and women's work: Phylogenetic cross-cultural analysis. American Journal of Physical Anthropology 110:27–45.

Hopkin, S. 1997. Biology of the Springtails (Insecta: Collembola). Oxford University Press, Oxford.

Hormiga, G., N. Scharff, and J. A. Coddington. 2000. The phylogenetic basis of sexual size dimorphism in orb-weaving spiders (Araneae, Orbiculariae). Systematic Biology 49:435–62.

Houde, A. 2001. Sex roles, ornaments, and evolutionary explanation. Proceedings of the National Academy of Sciences of the United States of America 98:12857–59.

Howell, F. G., and R. D. Ellender. 1984. Observations on growth and diet of *Argiope aurantia* Lucas (Araneidae) in a successional habitat. Journal of Arachnology 12:29–36.

Howell, W. M., and R. L. Jenkins. 2004. Spiders of the Eastern United States. Pearson Education, Boston.

Huber, B. 2005. Sexual selection research on spiders: Progress and biases. Biological Reviews 80:363–65.

Inkpen, S. A., and M. W. Foellmer. 2010. Sex-specific foraging behaviours and growth rates in juveniles contribute to the development of extreme sexual size dimorphism in a spider. The Open Ecology Journal 3:59–70.

Jaccarini, V., L. Agius, P. J. Schembri, and M. Rizzo. 1983. Sex determination and larval sexual interaction in *Bonellia viridis* Rolando (Echiura: Bonelliidae). Journal of Experimental Marine Biology and Ecology 66:25–40.

James, M. A., A. D. Ansell, and G. B. Curry. 1991. Functional morphology of the gonads of the articulate brachiopod *Terebratulina retusa*. Marine Biology 111:401–10.

Jarne, P., and J. R. Auld. 2006. Animals mix it up too: The distribution of self-fertilization among hermaphroditic animals. Evolution 60:1816–24.

Johnsgard, P. 1991. Bustards, Hemipods and Sandgrouse: Birds of Dry Places. Oxford University Press, Oxford, UK.

———. 1999. Earth, Water and Sky. A Naturalist's Stories and Sketches. University of Texas Press, Austin, Texas.

Johnson, J. C. 2001. Sexual cannibalism in fishing spiders (*Dolomedes triton*): An evaluation of two explanations for female aggression towards potential mates. Animal Behaviour 61:905–14.

Johnson, K. L., and L. G. Tassinary. 2007. Compatibility of basic social perceptions determines perceived attractiveness. Proceedings of the National Academy of Sciences of the United States of America 104:5246–51.

Jones, E. C. 1963. *Tremoctopus violaceus* uses *Physalia* tentacles as weapons. Science 139:764–66.

Jordan-Young, R. M. 2010. Brain Storm: The Flaws in the Science of Sex Differences. Harvard University Press, Cambridge, Massachusetts.

Jormalainen, V., S. Merilaita, and J. Tuomi. 1995. Differential predation on sexes affects colour polymorphism of the isopod *Idotea baltica* (Pallas). Biological Journal of the Linnean Society 55:45–68.

Kaiser, V. B., and H. Ellegran. 2006. Nonrandom distribution of genes with sex-biased expression in the chicken genome. Evolution 60:1945–51.

Kappeler, P. M., and C. P. van Schaik, eds. 2004. Sexual Selection in Primates: New and Comparative Perspectives. Cambridge University Press, Cambridge.

Kasumovic, M. M., M. J. Bruce, M. E. Herberstein, and M.C.B. Andrade. 2007. Risky mate search and mate preference in the golden orb-web spider (*Nephila plumipes*). Behavioral Ecology 18:189–95.

Katz, S., W. Klepal, and M. Bright. 2011. The *Osedax* trophosome: Organization and ultrastructure. Biol. Bull. 220:128–39.

Keenleyside, M.H.A. 1991. Cichlid Fishes. Behaviour, Ecology and Evolution. Chapman & Hall, London.

Kelber, A., M. Vorobyyev, and D. Osorio. 2003. Animal color vision—behavioural tests and physiological concepts. Biological Review 78:81–118.

Kelly, M. W., and E. Sanford. 2010. The evolution of mating systems in barnacles. Journal of Experimental Marine Biology and Ecology 392:37–45.

Kent, M. L., K. B. Andree, J. L. Bartholomew, M. El-Matbouli, S. S. Desser, R. H. Devlin, S. W. Feist, R. P. Hedrick, R. W. Hoffmann, J. Khattra, S. L. Hallett, R.J.G. Lester, M. Longshaw, O. Palenzuela, M. E. Siddall, and C. Xiao. 2001. Recent advances in our knowledge of the Myxozoa. The Journal of Eukaryotic Microbiology 48:395–413.

Kimball, R. T., and J. D. Ligon. 1999. Evolution of avian plumage dichromatism from a proximate perspective. The American Naturalist 154:182–93.

King, R. B. 2009. Sexual dimorphism in snake tail length: sexual selection, natural selection, or morphological constraint? Biological Journal of the Linnean Society 38:133–54.

Kraak, S., T. Bakker, and B. Mundwiler. 1999. Sexual selection in sticklebacks in the field: Correlates of reproductive, mating and paternal success. Behavioral Ecology 10:696–706.

Kraus, F. 2008. Remarkable case of anuran sexual size dimorphism: *Platymantis rhipiphalcus* is a junior synonym of *Platymantis boulengeri*. Journal of Herpetology 42:637–44.

Kristensen, R. M. 2002. An introduction to Loricifera, Cycliophora, and micrognathozoa. Integrative and Comparative Biology 42:641–51.

Kruuk, L. E., T. H. Clutton-Brock, K. E. Rose, and F. E. Guiness. 1999. Early determinants of lifetime reproductive success differ between the sexes in red deer. Proceedings of the Royal Society B 266:1655–61.

Kunte, K. 2008. Mimetic butterflies support Wallace's model of sexual dimorphism. Proceedings of the Royal Society B 275:1617–24.

Kuntner, M., and J. A. Coddington. 2009. Discovery of the largest orbweaving spider species: The evolution of gigantism in *Nephila*. PLoS One 4(10), e7516. doi: 10.1371/journal.pone.0007516.

Kupfer, A. 2007. Sexual size dimorphism in amphibians: An overview. Pp. 50–59 *in* D. J. Fairbairn, W. U. Blanckenhorn, and T. Székely, eds. Sex, Size and Gender Roles. Oxford University Press, Oxford.

Lamprell, K. L., and J. M. Healy. 2001. Scaphopoda. Pp. 85–128 *in* A. Wells and W.W.K. Houston, eds. Zoological Catalogue of Australia. CSIRO, Melbourne, Australia.

Land, M. F., and D.-E. Nilsson. 2004. Animal Eyes. Oxford University Press, Oxford, UK.

Landa, J., P. Pereda, R. Duarte, and M. Azevedo. 2001. Growth of anglerfish (*Lophius piscatorius* and *L. budegassa*) in Atlantic Iberian waters. Fisheries Research 51:363–76.

Laptikhovsky, V., and E. A. Salman. 2003. On reproductive strategies of the epipelagic octopods of the superfamily Argonautoidea (Cephalopoda: Octopoda). Marine Biology 142:321–26.

Laughlin, K. 2005. Management and control of egg size. The Poultry Site, http://www.thepoultrysite.com/articles/460/management-and-control-of-egg-size. Accessed February 24, 2009.

LeBas, N. R., and N. J. Marshall. 2000. The role of colour in signaling and male choice in the agamid lizard *Ctenophorus ornatus*. Proceedings of the Royal Society B 267:445–52.

Le Boeuf, B. J. 1994. Variation in the diving pattern of northern elephant seals with age, mass, sex and reproductive condition. Pp. 237–52 *in* B. J. Le Boeuf and R. M. Laws, eds. Elephant Seals: Population Ecology, Behavior, and Physiology. University of California Press, Berkeley, California.

Le Boeuf, B. J., D. E. Crocker, D. P. Costa, S. B. Blackwell, P. M. Webb, and D. S. Houser. 2000. Foraging ecology of northern elephant seals. Ecological Monographs 70:353–82.

Le Boeuf, B. J., and R. M. Laws. 1994a. Elephant seals: An introduction to the genus. Pp. 1–26 *in* B. J. Le Boeuf and R. M. Laws, eds. Elephant Seals: Population Ecology, Behavior, and Physiology. University of California Press, Berkeley, California.

———, eds. 1994b. Elephant Seals: Population Ecology, Behavior, and Physiology. University of California Press, Berkeley, California.

Le Boeuf, B. J., and S. Mesnick. 1991. Sexual behaviour of male northern elephant seals. I. Lethal injuries to adult females. Behaviour 116:1–26.

Le Boeuf, B. J., P. Morris, and J. Reiter. 1994. Juvenile survivorship of northern elephant seals. Pp. 121–36 *in* B. J. Le Boeuf and R. M. Laws, eds. Elephant

Seals: Population Ecology, Behavior, and Physiology. University of California Press, Berkeley, California.

Le Boeuf, B. J., and J. Reiter. 1988. Lifetime reproductive success in northern elephant seals. Pp. 344–62 *in* T. H. Clutton-Brock, ed. Reproductive Success. Studies of Variation in Contrasting Breeding Systems. University of Chicago Press, Chicago.

Levitan, D. R. 1996. Effects of gamete traits on fertilization in the sea and the evolution of sexual dimorphism. Nature 382:153–55.

Levitan, D. R., and C. Petersen. 1995. Sperm limitation in the sea. Trends in Ecology and Evolution 10:228–31.

Lewis, C., and T.A.F. Long. 2005. Courtship and reproduction in *Carybdea sivickisi* (Cnidaria: Cubozoa). Marine Biology 147:477–83.

Lewis, R., T. C. O'Connell, M. Lewis, C. Campagna, and A. R. Hoelzell. 2006. Sex-specific foraging strategies and resource partitioning in the southern elephant seal (*Mirounga leonina*). Proceedings of the Royal Society B 273:2901–7.

Li, J., Z. Zhang, F. Liu, Q. Liu, W. Gan, J. Chen, M.L.M. Lim, and D. Li. 2008. UVB-based mate-choice cues used by females in the jumping spider *Phintella vittata*. Current Biology 18:1–5.

Lima, M., and E. Páez. 1995. Growth and reproductive patterns in the South American fur seal. Journal of Mammalogy 76:1249–55.

Lindenfors, P. 2002. Sexually antagonistic selection on primate size. Journal of Evolutionary Biology 15:595–607.

Lindenfors, P., J. L. Gittleman, and K. Jones. 2007. Sexual size dimorphism in mammals. Pp. 16–26 *in* D. J. Fairbairn, W. U. Blanckenhorn, and T. Székely, eds. Sex, Size and Gender Roles. Evolutionary Studies of Sexual Size Dimorphism. Oxford University Press, Oxford, UK.

Lindenfors, P., and B. S. Tullberg. 1998. Phylogenetic analyses of primate size evolution: The consequences of sexual selection. Biological Journal of the Linnean Society 64:413–47.

Lindenfors, P., B. Tullberg, and M. Biuw. 2002. Phylogenetic analysis of sexual selection and sexual size dimorphism in pinnipeds. Behavioral Ecology and Sociobiology 52:188–93.

Lislevand, T., J. Figuerola, and T. Székely. 2007. Avian body sizes in relation to fecundity, mating system, display behavior, and resource sharing. Ecology 88:1605.

Lockley, T. C., and O. P. Young. 1993. Survivability of overwintering *Argiope aurantia* (Araneidae) egg cases, with an annotated list of associated arthropods. Journal of Arachnology 21:50–54.

Lombardi, J. 1998. Comparative Vertebrate Reproduction. Kluwer Academic Publishers, Norwell, Massachusetts.

Long, J. A., K. Trinajstic, G. Young, and T. Senden. 2008. Live birth in the Devonian period. Nature 453:650–52.

Lowerre-Barbieri, S. 2009. Reproduction in relation to conservation and exploitation of marine fishes. Pp. 371–94 *in* B.G.M. Jamieson, ed. Reproductive Biology and Phylogeny of Fishes (Agnathans and Bony Fishes). Part B. Science Publishers, Enfield, New Hampshire.

Loyan, A., M. S. Jalme, and G. Sorci. 2005. Inter- and intra-sexual selection for multiple traits in the peacock (*Pavo cristatus*). Ethology 111:810–20.

Lu, C. C. 2001. Cephalopoda. Pp. 129-308 *in* A. Wells and W.W.K. Houston, eds. Zoological Catalogue of Australia. Volume 17.2. Mollusca: Aplacophora, Polyplacorphora, Scaphopoda, Cephalopoda. CSIRO, Melbourne, Australia.

Lundsten, L., C. K. Paull, K. L. Schlining, M. McGann, and W. Ussler. 2010. Biological characterization of a whale-fall near Vancouver Island, British Columbia, Canada. Deep-Sea Research Part I—Oceanographic Research Papers 57:918–22.

Lyon, M. F. 1994. Evolution of mammalian sex-chromosomes. Pp. 381–96 *in* A. V. Short and E. Balaban, eds. The Differences between the Sexes. Cambridge University Press, Cambridge, UK.

Maan, M. E., and M. Taborsky. 2008. Sexual conflict over breeding substrate causes female expulsion and offspring loss in a cichlid fish. Behavioral Ecology 19:302–8.

Mackenzie, A., J. D. Reynolds, V. J. Brown, and W. J. Sutherland. 1995. Variation in male mating success on leks. The American Naturalist 145:633–52.

Magaña, M., J. C. Alonso, C. A. Martín, L. M. Bautista, and B. Martín. 2010. Nest-site selection by great bustards *Otis tarda* suggests a trade-off between concealment and visibility. IBIS 152:77–89.

Maggenti, A. R. 1981. Nematodes: Development as plant parasites. Annual Reviews of Microbiology 35:135–54.

Mangold, K. M., M. Vecchione, and R. E. Young. 2010a. Tremoctopodidae Tyron 1897. *Tremoctopus* Chiaie 1830. Blanket octopus. Version 15 August 2010. http://tolweb.org/Tremoctopus/20202/2010.08.15 *in* The Tree of Life Web Project, http://Tolweb.org/. Accessed August 25, 2011.

———. 2010b. Ocythoidae Gray 1849. *Ocythoe tuberculata* Rafinesque 1814. Version 15 August 2010 (under construction). http://toweb.org/Ocythoe

tuberculata/ 20205/2010.08.15 *in* The Tree of Life Web Project, http://
Tolweb.org/. Accessed August 25, 2010.

Mank, J. E., and J. C. Avise. 2006. Comparative phylogenetic analysis of male
alternative reproductive tactics in ray-finned fishes. Evolution 60:1311–16.

Marchand, B., and G. Vassilliades. 1982. *Mediorhynchus mattei* sp. N. (Acanthocephala, Giganthorhynchidae) from *Tockus erthrorhynchus* (Aves), the
red-beaked hornbill in West Africa. Journal of Parasitology 68:1142–45.

Marin, I., and B. S. Baker. 1998. The evolutionary dynamics of sex determination. Science 281:1990–94.

Marshall Graves, J. A. 1994. Mammalian sex-determining genes. Pp. 397–418
in R. V. Short and E. Balaban, eds. The Differences between the Sexes. Cambridge University Press, Cambridge.

Martín, C. A., J. C. Alonso, J. Alonso, C. Pitra, and D. Lieckfeldt. 2001a.
Great bustard population structure in central Spain: Concordant results
from genetic analysis and dispersal study. Proceedings of the Royal Society B
269:119–25.

Martín, C. A., J. C. Alonso, J. A. Alonso, C. Palacín, M. Magaña, and B. Martín.
2007. Sex-biased juvenile survival in a bird with extreme size dimorphism, the
great bustard *Otis tarda*. Journal of Avian Biology 38:335–46.

Martín, C. A., J. C. Alonso, M. B. Morales, and S. J. Lane. 2001b. Seasonal
movements of male great bustards in central Spain. Journal of Field Orni-
thology 72:504–8.

Martinez, C. 1988. Size and sex composition of great bustard (*Otis tarda*) flocks
in Villafafila, northwest Spain. Ardeoloa 35:125–33.

McConnell, B. J., C. Chambers, and M. A. Fedak. 1992. Foraging ecology of
southern elephant seals in relation to the bathymetry and productivity of the
Southern Ocean. Antarctic Science 4:393–98.

McDermott, J. J. 2001. Symbionts of the hermit crab *Pagurus longicarpus* Say,
1817 (Decapoda: Anomura): New observations from New Jersey waters and
a review of all known relationships. Proceedings of the Biological Society of
Washington 114:624–39.

McDermott, J. J., and R. Gibson. 1993. *Carcinonemertes pinnotheridophila* sp.
nov. (Nemertea, Enopla, Carcinonemertidae) from the branchial chambers
of *Pinnixa chaetopterana* (Crustacea, Decapoda, Pinnotheridae): Description, incidence and biological relationships with the host. Hydrobiologia
299:57–80.

McDonald, B. I., and D. E. Crocker. 2006. Physiology and behavior influence lactation efficiency in northern elephant seals (*Mirounga angustirostris*). Physiological and Biochemical Zoology 79:484–96.

McFadien-Carter, M. 1979. Scaphopoda. Pp. 95–111 *in* A. C. Giese and J. S. Pearse, eds. Reproduction of Marine Invertebrates, Volume V. Molluscs: Pelecypods and Lesser Classes. Academic Press, New York.

McMahon, C. R., H. R. Burton, and M. N. Bester. 2000. Weaning mass and the future survival of juvenile southern elephant seals, *Mirounga leonina*, at Macquarie Island. Antarctic Science 12:149–53.

McReynolds, C. N. 2000. The impact of habitat features on web features and prey capture of *Argiope aurantia* (Araneae, Araneidae). Journal of Arachnology 28:169–79.

Mealey, L. 2000. Sex Differences: Developmental and Evolutionary Strategies. Academic Press, San Diego, California.

Meidl, P. 1999. Microsatellite analysis of alternative mating tactics in *Lamprologus callipterus*. Diplomarbeit, am Konrad Lorenz-Institut für Vergleichende Verhaltensforschung, University of Vienna, Vienna, Austria.

Meraz, L. C., Y. Henaut, and M. A. Elgar. 2012. Effects of male size and female dispersion on male mate-locating success in *Nephila clavipes*. Journal of Ethology 30:93–100.

Mesa, A., P. García-Novo, and D. dos Santos. 2002. X_1X_2O (male)–$X_1X_1X_2X_2$ (female) chromosomal sex determining mechanism in the cricket *Cicloptyloides americanus* (Orthoptera, Grylloidea, Mogoplistidae). Journal of Orthoptera Research 11:87–90.

Mesnick, S. L., and B. J. Le Boeuf. 1991. Sexual behavior of male northern elephant seals: II. Female response to potentially injurious encounters. Behaviour 117:262–80.

Messina, F., and C. W. Fox. 2001. Offspring size and number. Pp. 113–27 *in* C. W. Fox, D. A. Roff, and D. J. Fairbairn, eds. Evolutionary Ecology. Concepts and Case Studies. Oxford University Press, New York.

Metaxas, A., and N. E. Kelly. 2010. Do larval supply and recruitment vary among chemosynthetic environments of the deep sea? PloS One 5(7): e11646. doi: 10.1371/journal.pone.0011646.

Millar, N. P., D. N. Reznick, M. T. Kinnison, and A. P. Hendry. 2006. Disentangling the selective factors that act on male colour in wild guppies. Oikos 113:1–12.

Miller, J. A. 2007. Repeated evolution of male self-sacrifice behavior in spiders correlated with genital mutilation. Evolution 61:1601–13.

Mindell, D. P., J. W. Brown, and J. Harshman. 2008. Tree of Life Project: Neoaves. Version 27 June 2008 (under construction). http://tolweb.org/ Neoaves/26305/2008.06.27 *in* the Tree of Life Project, http://tolweb.org/. Accessed March 24, 2010.

Minelli, A., D. Foddai, A. Pereira, and J.G.E. Lewis. 2000. The evolution of segmentation of centipede trunk and appendages. Journal of Zoological Systematics and Evolutionary Research 38:103–17.

Ming, R., and P. H. Moore. 2007. Genomics of sex chromosomes. Current Opinion in Plant Biology 10:123–30.

Modig, A. O. 1996. Effects of body size and harem size on male reproductive behaviour in the southern elephant seal. Animal Behaviour 51:1295–1306.

Moore, A. J. 1990. The evolution of sexual dimorphism by sexual selection: The separate effects of intrasexual selection and intersexual selection. Evolution 44:315–31.

Morales, M. B., J. C. Alonso, and J. Alonso. 2002. Annual productivity and individual female reproductive success in a great bustard *Otis tarda* population. Ibis 144:293–300.

Morales, M. B., J. C. Alonso, J. A. Alonso, and E. Martín. 2000. Migration patterns in male great bustards (*Otis tarda*). Auk 117:493–98.

Morales, M. B., J. C. Alonso, C. Martin, E. Martin, and J. Alonso. 2003. Male sexual display and attractiveness in the great bustard *Otis tarda*: The role of body condition. Journal of Ethology 21:51–56.

Morales, M. B., F. Jiguet, and B. Arroyo. 2001. Exploded leks: What bustards can teach us. Ardeola 48:85–98.

Morgado, R., and F. Moreira. 2000. Seasonal population dynamics, nest site selection, sex-ratio and clutch size of the Great Bustard *Otis tarda* in two adjacent lekking areas. Ardeola 47:237–46.

Morrow, E. H. 2004. How the sperm lost its tail: The evolution of aflagellate sperm. Biological Reviews 79:795–814.

Moya-Laraño, J., D. Vinkovic, C. M. Allard, and M. W. Foellmer. 2009. Optimal climbing speed explains the evolution of extreme sexual size dimorphism in spiders. Journal of Evolutionary Biology 22:954–63.

Moya-Laraño, J., D. Vinkovic, E. De Mas, G. Corcobado, and E. Moreni. 2008. Morphological Evolution of Spiders Predicted by Pendulum Mechanics. PLoS One 3(3): e1841. doi:10.1371/journal.pone.0001841.

Nagy, S. 2007. International Single Species Action Plan for the Western Pale-arctic Population of Great Bustard *Otis tarda tarda*. http://ec.europa.eu/environment/nature/conservation/wildbirds/action_plans/docs/otis_tarda.pdf. Accessed October 2012.

Neilsen, C. 2001. Animal Evolution: Interrelationships of the Living Phyla. 2nd edition. Oxford University Press, Oxford, UK.

Nesis, K. N. 1987. Cephalopods of the World. Squids, Cuttlefishes, Octopuses and Allies. TFH Publications, Neptune City, New Jersey.

Nessler, S. H., G. Uhl, and J. M. Schneider. 2009. Scent of a woman—the effect of female presence on sexual cannibalism in an orb-weaving spider (Araneae: Araneidae). Ethology 115:633–40.

Neuhaus, B., and R. P. Higgins. 2002. Ultrastructure, biology and phylogenetic relationships of the Kinorhyncha. Integrative and Comparative Biology 42:619–32.

Norman, M. D., D. Paul, J. Finn, and T. Tregenza. 2002. First encounter with a live male blanket octopus: The world's most sexually size-dimorphic large animal. New Zealand Journal of Marine and Freshwater Research 36:733–36.

Nyffeler, M., D. Dean, and W. Sterling. 1987. Feeding ecology of the orb-weaving spider *Argiope aurantia* (Araneae, Araneidae) in a cotton agroeco-system. Entomophaga 32:367–76.

Obst, M., and P. Funch. 2003. Dwarf male of *Symbion pandora* (Cycliophora). Journal of Morphology 255:261–78.

O'Dor, R. K., and M. J. Wells. 1987. Energy and nutrient flow. Pp. 109–33 *in* P. R. Boyle, ed. Cephalopod Life Cycles. Volume II. Comparative Reviews. Academic Press, London.

Ohtsuka, S., and R. Huys. 2001. Sexual dimorphism in calanoid copepods: Morphology and function. Hydrobiologia 453/454:441–66.

Oliver, B., and M. Parisi. 2004. Battle of the Xs. BioEssays 26:543–48.

Oliver, J. C., K. A. Robertson, and A. Monteiro. 2009. Accommodating natural and sexual selection in butterfly wing pattern evolution. Proceedings Biological Science 276:2369–75.

O'Loughlin, P. M. 2001. The occurrence and role of genital papilla in holo-thurian reproduction. Pp. 363–75 *in* M. Barker, ed. Echinoderms 2000. Proceedings of the 10th International Conference, Dundedin, January 31–February 4, 2000. Swets & Zeitlinger, Lisse, The Netherlands.

O'Shea, S. 1999. The marine fauna of New Zealand: Octopoda (Mollusca: Cephalopoda). NIWA Biodiversity Memoir 112:1–280.

O'Shea, S. 2010. The giant octopus *Haliphron atlanticus* (Mollusca: Octopoda) in New Zealand waters. New Zealand Journal of Zoology 31:9–13.

Ostrovsky, A. N., and J. S. Porter. 2011. Pattern of occurrence of supraneural coelomopores and intertentacular organs in Gymnolaemata (Bryozoa) and its evolutionary implications. Zoomorphology 130:1–15.

Ota, K., M. Kohda, and T. Sato. 2010. Unusual allometry for sexual size dimorphism in a cichlid where males are extremely larger than females. Journal of Bioscience 35:257–65.

Ouellet, P., Y. Lambert, and I. Bérubé. 2001. Cod egg characteristics and viability in relation to low temperature and maternal nutritional condition. ICES Journal of Marine Science 58:672–86.

Owens, I.P.F., and R. V. Short. 1995. Hormonal basis of sexual dimorphism in birds: Implications for new theories of sexual selection. Trends in Ecology and Evolution 10:44–47.

Palacín, C., J. C. Alonso, J. A. Alonso, C. A. Martín, M. Magaña, and B. Martin. 2009. Differential migration by sex in the great bustard: Possible consequences of an extreme sexual size dimorphism. Ethology 115:617–26.

Pan, L., C. Ober, and M. Abney. 2007. Heritability estimation of sex-specific effects on human quantitative traits. Genetic Epidemiology 31:338–47.

Parker, G. A. 1992. The evolution of sexual size dimorphism in fish. Journal of Fish Biology 41 (Supplement B):1–20.

Parker, G. A., R. R. Baker, and V.G.F. Smith. 1972. The origin and evolution of gamete dimorphism and the male-female phenomenon. Journal of Theoretical Biology 36:181–98.

Parr, A. E. 1930. On the probable identity, life history and anatomy of the free-living and attached males of the ceratioid fishes. Copeia 1930:129–35.

Partecke, J., A. von Haeseller, and M. Wikelski. 2002. Territory establishment in lekking marine iguanas, *Amblyrhynchus cristatus*: Support for the hotshot mechanism. Behavioral Ecology and Sociobiology 51:579–87.

Pearse, J. S. 1979. Polyplacophora. Pp. 27–85 *in* A. C. Giese and J. S. Pearse, eds. Reproduction of Marine Invertebrates. Volume V. Molluscs: Pelecypods and Lesser Classes. Academic Press, New York.

Pearse, V. B., and O. Voigt. 2007. Field biology of placozoans (*Trichoplax*): Distribution, diversity, biotic interactions. Journal of Integrative and Comparative Biology 47:677–92.

Pearson, D., R. Shine, and A. Williams. 2002. Geographic variation in sexual size dimorphism within a single snake species (*Morelia spilota*, Pythonidae). Oecologia 131:418–26.

Pechenik, J. A. 2005. Biology of the Invertebrates, 5th edition. McGraw-Hill, New York.

Pecl, G. T., and N. A. Moltschaniwskyj. 2006. Life history of a short-lived squid (*Sepioteuthis australis*): Resource allocation as a function of size, growth, maturation, and hatching season. ICES Journal of Marine Science 63:995–1004.

Peichi, L., G. Behrmann, and R.H.H. Kronègers. 2001. For whales and seals the ocean is not blue: A visual pigment loss in marine mammals. European Journal of Neuroscience 13:1520–28.

Pelletier, F., and M. Festa-Bianchet. 2006. Sexual selection and social rank in bighorn rams. Animal Behaviour 71:641–55.

Peters, R. H. 1983. The Ecological Implications of Body Size. Cambridge University Press, Cambridge, UK.

Petraits, P. S. 1985. Digametic sex determination in the marine polychaete, *Capitella capitata* (species type I). Heredity 55:151–56.

Pfennig, D. W., W. R. Harcombe, and K. S. Pfennig. 2001. Frequency-dependent Batesian mimicry. Nature 410:323.

Philippe, H., N. Lartillot, and H. Brinkmann. 2005. Multigene analyses of bilaterian animals corroborate the monophyly of Ecdysozoa, Lophotrochozoa, and Protostomia. Molecular Biology and Evolution 22:1246–53.

Pianka, E. R., and L. J. Vitt. 2003. Lizards. Windows to the Evolution of Diversity. University of California Press, Berkeley and Los Angeles, California.

Pietsch, T. W. 2005. Dimorphism, parasitism, and sex revisited: Modes of reproduction among deep-sea ceratioid anglerfishes (Teleostei: Lophiiformes). Ichthyological Research 52:207–36.

———. 2009. Oceanic Anglerfishes. Extraordinary Diversity in the Deep Sea. University of California Press, Berkeley and Los Angeles, California.

Pietsch, T. W., and D. B. Grobecker. 1987. Frogfishes of the World: Systematics, Zoogeography and Behavioral Ecology. Stanford University Press, Stanford, California.

Pietsch, T. W., and C. P. Kenaley. 2007. Ceratioidei. Seadevils, Devilfishes, Deep-sea Anglerfishes. Version 02, October 2007 (under construction), http://tolweb.org/Ceratioidei/22000/2007.10.02 *in* The Tree of Life Web Project, http://tolweb.org/. Accessed September 5, 2011.

Pistorius, P. A., M. N. Bester, G.J.G. Hofmeyr, S. P. Kirkman, and F. E. Taylor. 2008. Seasonal survival and the relative cost of first reproduction in adult female southern elephant seals. Journal of Mammalogy 89:567–74.

Pitnick, S., D. J. Hosken, and T. R. Birkhead. 2009. Sperm morphological diversity. Pp. 69–149 *in* T. R. Birkhead, D. J. Hosken, and S. Pitnick, eds. Sperm Biology: An Evolutionary Perspective. Elsevier, Burlington, Massachusetts and San Diego, California.

Pitra, C. D., S. Frahnert, and J. Fickel. 2002. Phylogenetic relationships and ancestral areas of the bustards (Gruiformes: Otididae), inferred from mitochondrial DNA and nuclear intron sequences. Molecular Phylogenetic and Evolution 23:63–74.

Pizzari, T., and G. A. Parker. 2009. Sperm competition and sperm phenotype. Pp. 207–45 *in* T. R. Birkhead, D. J. Hosken, and S. Pitnick, eds. Sperm Biology: An Evolutionary Perspective. Elsevier, Burlington, Massachusetts and San Diego, California.

Plavcan, J. M. 2004. Sexual selection, measures of sexual selection and sexual dimorphism in primates. Pp. 230–52 *in* P. M. Kappeler and C. P. van Schaik, eds. Sexual Selection in Primates: New and Comparative Perspectives. Cambridge University Press, Cambridge, UK.

Post, E., R. Langvatin, M. C. Forchhammer, and N. C. Stenseth. 1999. Environmental variation shapes sexual dimorphism in red deer. Proceedings of the National Academy of Sciences of the United States of America 96:4467–71.

Poulin, R. 1996. Sexual size dimorphism and transition to parasitism in copepods. Evolution 50:2520–23.

———. 1997. Covariation of sexual size dimorphism and adult sex ratio in parasitic nematodes. Biological Journal of the Linnean Society 62:567–80.

Poulin, R., and S. Morand. 2000. Testes size, body size and male-male competition in acanthocephalan parasites. Journal of Zoology London 250:551–58.

Prenter, J., R. W. Elwood, and W. I. Montgomery. 1999. Sexual size dimorphism and reproductive investment by female spiders: a comparative analysis. Evolution 53:1987–94.

Prenter, J., B. G. Fanson, and P. W. Taylor. 2012. Whole-organism performance and repeatability of locomotion on inclines in spiders. Animal Behaviour 83:1195–1201.

Prenter, J., D. Pérez-Staples, and P. W. Taylor. 2010. The effects of morphology and substrate diameter on climbing and locomotor performance in male spiders. Functional Ecology 24:400–8.

Preziosi, R. F., and D. J. Fairbairn. 2000. Lifetime selection on adult body size and components of body size in a waterstrider: Opposing selection and maintenance of sexual size dimorphism. Evolution 54:558–66.

Prokop, Z. M., L. Michalczyk, S. M. Drobniak, M. Herdegen, and J. Radwan. 2012. Meta-analysis suggests choosy females get sexy sons more than "good genes." Evolution 66:1–9.

Quirk, J. 2006. Sperm Are from Men. Eggs Are from Women. Running Press, Philadelphia, Pennsylvania.

Raihani, G., T. Székely, M. A. Serrano-Meneses, C. Pitra, and P. Goriup. 2006. The influence of sexual selection and male agility on sexual size dimorphism in bustards (Otididae). Animal Behaviour 71:833–38.

Ralls, K. 1976. Mammals in which females are larger than males. The Quarterly Review of Biology 51:245–76.

———. 1977. Sexual dimorphism in mammals: Avian models unanswered questions. American Naturalist 111:917–38.

Regan, C. T. 1925. Dwarfed males parasitic on the females in oceanic angler-fishes (Pediculati, Ceratioidea). Proceedings of the Royal Society B 97:386–400.

Reiss, M. J. 1989. The Allometry of Growth and Reproduction. Cambridge University Press, Cambridge.

Reunov, A. A. 2005. Problem of terminology in characteristics of spermatozoa of metazoa. Russian Journal of Developmental Biology 36:335–51.

Ricci, C., G. Melone, and C. Sotgia. 1993. Old and new data on Seisonidea (Rotifera). Hydrobiologia 255/256:495–511.

Ridley, M. 1993. The Red Queen. Sex and the Evolution of Human Nature. Penguin Books, New York.

Robinson, M. H., and B. R. Robinson, eds. 1980. Pacific Insects Monograph 36. Department of Entomology, Bishop Museum, Honolulu, Hawaii.

Rocha, F., A. Guerra, and A. F. Gonzalez. 2001. A review of reproductive strategies in cephalopods. Biological Review 76:291–304.

Roe, P. 1993. Aspects of the biology of *Pantinonemertes californiensis*, a high intertidal nemertean. Hydrobiologia 266:29–44.

Roff, D., A. 1990. The evolution of flightlessness in insects. Ecological Monographs 60:389–421.

———. 1991. Life history consequences of bioenergetic and biomechanical constraints on migration. American Zoologist 31:205-215.

———. 1992. The Evolution of Life Histories. Chapman and Hall, New York.

Roulin, A. 1999. Nonrandom pairing by male barn owls (*Tyto alba*) with respect to a female plumage trait. Behavioral Ecology 10:688–95.

Rouse, G. W., S. K. Goffredi, S. B. Johnson, and R. C. Vrijenhoek. 2011. Not whale-fall specialists, *Osedax* worms also consume fish bones. Biology Letters 7:736–39.

Rouse, G. W., S. K. Goffredi, and R. C. Vrijenhoek. 2004. *Osedax*: Bone-eating marine worms with dwarf males. Science 305:668–71.

Rouse, G. W., N. G. Wilson, S. K. Goffredi, S. B. Johnson, T. Smart, C. Widmer, C. M. Young, and R. C. Vrijenhoek. 2009. Spawning and development in *Osedax* boneworms (Siboglinidae, Annelida). Marine Biology 156:395–405.

Rouse, G. W., K. Worsaae, S. B. Johnson, W. J. Jones, and R. C. Vrijenhoek. 2008. Acquisition of dwarf male "harems" by recently settled females of *Osedax roseus* n. sp. (Siboglinidae; Annelida). Biological Bulletin 214:67–82.

Ruff, C. 2002. Variation in human body size and shape. Annual Review of Anthropology 31:211–32.

Russo, V.E.A., R. A. Martienssen, and A. D. Riggs, eds. 1996. Epigenetic Mechanisms of Gene Regulation. Cold Spring Harbor Laboratory Press, Plainview, New York.

Rutowski, R. L. 2003. Visual ecology of adult butterflies. Pp. 9–25 *in* C. L. Boggs, W. B. Watt, and P. R. Ehrlich, eds. Butterflies: Ecology and Evolution Take Flight. University of Chicago Press, Chicago, London.

Sanvito, S., F. Galimberti, and E. H. Miller. 2007a. Having a big nose: Structure, ontogeny, and function of the elephant seal proboscis. Canadian Journal of Zoology 85:207–20.

———. 2007b. Vocal signaling of male southern elephant seals is honest but imprecise. Animal Behaviour 73:287–99.

———. 2008. Development of aggressive vocalizations in male southern elephant seals (*Mirounga leonina*): Maturation or learning? Behaviour 14:137–70.

Sastry, A. N. 1979. Pelecypoda (excluding Ostreidae). Pp. 113–292 *in* A. C. Giese and J. S. Pearse, eds. Reproduction of Marine Invertebrates, Volume V. Molluscs: Pelecypods and Lesser Classes. Academic Press, New York.

Sato, T. 1994. Active accumulation of spawning substrate: A determinant of extreme polygyny in a shell-brooding cichlid fish. Animal Behaviour 48:669–78.

Sato, T., M. Hirose, M. Taborsky, and S. Kimura. 2004. Size-dependent male alternative reproductive tactics in the shell-brooding fish *Lamprologus callipterus* in Lake Tanganyika. Ethology 110:49–62.

Scherer, G., and M. Schmid, eds. 2001. Genes and Mechanisms of Vertebrate Sex Determination. Birkhäuser Verlag, Basel, Switzerland.

Schmidt-Nielsen, K. 1984. Scaling: Why Is Animal Size So Important? Cambridge University Press, Cambridge, UK.

Schmidt-Rhaesa, A. 2002. Two dimensions of biodiversity research exemplified by Nematomorpha and Gastrotricha. Integrative and Comparative Biology 42:633–40.

Schneider, J. M., and M. A. Elgar. 2001. Sexual cannibalism and sperm competition in the golden orb-web spider Nephila plumipes (Araneoidea): Female and male perspectives. Behavioral Ecology 12:547–52.

Schneider, J. M., S. Gilberg, L. Fromhage, and G. Uhl. 2006. Sexual conflict over copulation duration in a cannibalistic spider. Animal Behaviour 71:781–88.

Schneider, J. M., M. E. Herberstein, F. C. de Crespigny, S. Ramamurthy, and M. A. Elgar. 2000. Sperm competition and small size advantage for males of the golden orb-web spider Nephila edulis. Journal of Evolutionary Biology 13:939–46.

Schütz, D., G. Pachler, E. Ripmeester, O. Goffinet, and M. Taborsky. 2010. Reproductive investment of giants and dwarfs: Specialized tactics in a cichlid fish with alternative male morphs. Functional Ecology 24:131–49.

Schütz, D., G. A. Parker, M. Taborsky, and T. Sato. 2006. An optimality approach to male and female body sizes in an extremely size-dimorphic cichlid fish. Evolutionary Ecology Research 8:1393–1408.

Schütz, D., and M. Taborsky. 2000. Giant males or dwarf females: What determines the extreme sexual size dimorphism in Lamprologus callipterus? Journal of Fish Biology 57:1254–65.

———. 2005. The influence of sexual selection and ecological constraints on an extreme sexual dimorphism in a cichlid. Animal Behaviour 70:539–49.

Schwabe, E. 2008. A summary of reports of abyssal and hadal Monoplacophora and Polyplacophora (Mollusca). Zootaxa 1866:2005–22.

Seifan, M., A. Gilad, K. Klass, and Y. L. Werner. 2009. Ontogenetically stable dimorphism in a lacertid lizard (Acanthodactylus boskianus) with tests of methodology and comments on life history. Biological Journal of the Linnean Society 97:275–88.

Setchell, J. M., and A. F. Dixson. 2001. Changes in the secondary sexual adornments of male mandrills (Mandrillus sphinx) are associated with gain and loss of alpha status. Hormones and Behavior 39:177–84.

Sharma, G. D., and L. J. Metz. 1976. Biology of the Collembola *Xenylla grisea* Axelson and *Lepidocyrtus cyaneus* f. *cinereus* Folsom. Ecological Entomology 1:209–12.

Shelly, T. E., and T. S. Whittier. 1997. Lek behavior of insects. Pp. 273–93 *in* J. C. Choe and B. J. Crespi, eds. The Evolution of Mating Systems in Insects and Arachnids. Cambridge University Press, Cambridge.

Shine, R. 1979. Sexual selection and sexual dimorphism in the Amphibia. Copeia 2:297–306.

Simmons, L. W. 2001. Sperm Competition and Its Evolutionary Consequences in the Insects. Princeton University Press, Princeton, New Jersey.

Simonini, R., F. Molinari, M. Pagliaia, I. Anasaloni, and D. Prevedelli. 2003. Karyotype and sex determination in *Dinophilus gyrociliatus* (Polychaeta: Dinophilidae). Marine Biology 142:441–45.

Sims, D. W. 2005. Differences in habitat selection and reproductive strategies of male and female sharks. Pp. 127–47 *in* K. E. Ruckstuhl and P. Neuhaus, eds. Sexual Segregation in Vertebrates. Ecology of the Two Sexes. Cambridge University Press, Cambridge, UK.

Sinervo, B., and C. Lively. 1996. The rock-paper-scissors game and the evolution of alternative male strategies. Nature 380:240–43.

Sivinski, J. 1978. Intrasexual aggression in the stick insects, *Diapheromera veliei* and *D. covilleae*, and sexual dimorphism in the Phasmatodea. Psyche 85:395–406.

Skinner, M. M., and B. Wood. 2006. The evolution of modern human life history: A paleontological perspective. Pp. 331–64 *in* K. Hawkes and R. R. Paine, eds. The Evolution of Human Life History. School of American Research Press, Santa Fe, New Mexico.

Skuse, D. 2006. Sexual dimorphism in cognition and behaviour: The role of X-linked genes. European Journal of Endochrinology 155:S99–106.

Slip, D. J., M. A. Hindell, and H. R. Burton. 1994. Diving behavior of southern elephant seals from Macquarie Island. Pp. 253–70 *in* B. J. Le Boeuf and R. M. Laws, eds. Elephant Seals: Population Ecology, Behavior, and Physiology. University of California Press, Berkeley, California.

Smith, D. G. 1972. The role of the epaulets in the red-winged blackbird (*Agelaius phoeniceus*) social system. Behaviour 41:251–68.

Stewart, B. S., and R. L. DeLong. 1994. Postbreeding foraging migrations of northern elephant seals. Pp. 290–309 *in* B. J. Le Boeuf and R. M. Laws, eds.

Elephant Seals: Population Ecology, Behavior, and Physiology. University of California Press, Berkeley, California.

Stockley, P., M. J. Gage, G. A. Parker, and A. P. Møller. 1997. Sperm competition in fishes: The evolution of testis size and ejaculate characteristics. The American Naturalist 149:933–54.

Stöhr, S. 2001. *Amphipholis linpneusti* n. sp., a sexually dimorphic amphiurid brittle star (Echinodermata: Ophiuroidea), epizoic on a spanagoid sea urchin. Pp. 317–22 *in* M. Barker, ed. Echinoderms 2000. A. A. Balkema, Swets and Zeitlinger, Lisse, The Netherlands.

Stricker, S. A., T. L. Smythe, L. Miller, and J. L. Norenburg. 2000. Comparative biology of oogenesis in nemertean worms. Acta Zoologica 82:213–30.

Striech, W. J., H. Litzbarski, B. Ludwig, and S. Ludwig. 2006. What triggers facultative winter migration of great bustard (*Otis tarda*) in Central Europe? European Journal of Wildlife Research 52:48–53.

Sturmbauer, C., E. Verheyen, and A. Meyer. 1994. Mitochondrial phylogeny of the Lamprologini, the major substrate spawning lineage of cichlid fishes from Lake Tanganyika in eastern Africa. Molecular Biology and Evolution 11:691–703.

Sydeman, W. J., and N. Nur. 1994. Life history strategies of female northern elephant seals. Pp. 137–53 *in* B. J. Le Boeuf and R. M. Laws, eds. Elephant Seals: Population Ecology, Behavior, and Physiology. University of California Press, Berkeley, California.

Székely, T., R. P. Freckleton, and J. D. Reynolds. 2004. Sexual selection explains Rensch's rule of size dimorphism in shorebirds. Proceedings of the National Academy of Sciences of the United States of America 101:12224–27.

Székely, T., T. Lislevand, and J. Figuerola. 2007. Sexual size dimorphism in birds. Pp. 27–37 *in* D. J. Fairbairn, W. U. Blanckenhorn, and T. Székely, eds. Sex, Size and Gender Roles. Evolutionary Studies of Sexual Size Dimorphism. Oxford University Press, Oxford.

Taborsky, M. 2001. The evolution of bourgeois, parasitic and cooperative reproductive behaviors in fishes. Journal of Heredity 92:100–10.

Talebizadeh, Z., S. D. Simon, and M. G. Butler. 2006. X chromosome gene expression in human tissues: Male and female comparisons. Genomics 88: 675–81.

Temereva, E. N., and V. V. Malakhov. 2001. The morphology of the phoronid *Phoronopsis harmeri*. Russian Journal of Marine Biology 27:21–30.

Thomas, R. F. 1977. Systematics, distribution, and biology of cephalopods of the genus *Tremoctopus* (Octopoda: Tremoctopodidae). Bulletin of Marine Science 27:353–92.

Thorsen, A., O. S. Kjesbu, H. J. Fyhn, and P. Solemdal. 1996. Physiological mechanisms of buoyancy in eggs from brackish water cod. Journal of Fish Biology 48:457–77.

Tolbert, W. W. 1975. Predator avoidance behaviors and web defensive structures in the orb weavers *Argiope aurantia* and *Argiope trifasciata* (Araneae, Araneidae). Psyche 82:29–52.

Tominaga, H., S. Nakamura, and M. Komatsu. 2004. Reproduction and development of the conspicuously dimorphic brittle star *Ophiodaphne formata* (Ophiuroidea). Biological Bulletin 206:25–34.

Tomlinson, J. T. 1969a. The burrowing barnacles (Cirripedia: Order Acrothoracica). Bulletin–the United States National Museum 296:169.

———. 1969b. Shell-burrowing barnacles. American Zoologist 9:837–40.

Tree of Life Web Project. 2002a. Animals. Metazoa. Version 01 January 2002 (temporary). http://tolweb.org/Animals/2374/2002.01.01 *in* The Tree of Life Web Project, http://tolweb.org/. Accessed October 28, 2012.

———. 2002b. Bilateria. Triploblasts, bilaterally symmetrical animals with three germ layers. Version 01 January 2002 (temporary). http://tolweb.org/Bilateria/2459/2002.01.01 *in* The Tree of Life Web Project, http://tolweb.org/. Accessed October 28, 2012.

———. 2002c. Deuterostomia. Version 01 January 2002 (temporary). http://tolweb.org/Deuterostomia/2466/2002.01.01 *in* The Tree of Life Web Project, http://tolweb.org/. Accessed October 28, 2012.

———. 2009. Cirripedia. Version 10 December 2009 (temporary). http://tolweb.org/Cirripedia/8127/2009.12.10 *in* The Tree of Life Web Project, http://tolweb.org/. Accessed September 30, 2011.

Uhl, G., and F. Vollrath. 1998. Little evidence for size-selective sexual cannibalism in two species of *Nephila* (Araneae). *Zoology* 101:101–6.

Uller, T., I. Pen, E. Wapstra, L. W. Beukeboom., and J. Komdeur. 2007. The evolution of sex ratios and sex-determining systems. Trends in Ecology and Evolution 22:292–97.

Urano, S., S. Yamaguchi, S. Yamato, S. Takahashi, and Y. Yusa. 2009. Evolution of dwarf males and a variety of sexual modes in barnacles: An ESS approach. Evolutionary Ecology Research 11:713–29.

Vahed, K., D. J. Parker, and J.D.J. Gilbert. 2011. Larger testes are associated with a higher level of polyandry, but a smaller ejaculate volume, across bush-cricket species (Tettigoniidae). Biology Letters 7:261–64

Vail, L. 1987. Reproduction in five species of crinoids at Lizard Island, Great Barrier Reef. Marine Biology 95:431–46.

Valian, V. 1999. Why So Slow? The Advancement of Women. MIT Press, Cambridge, Massachusetts.

Vandeputte, M., M. Dupont-Nivet, H. Chavanne, and B. Chatain. 2007. A polygenic hypothesis for sex determination in the European sea bass *Dicentrarchus labrax*. Genetics 176:1049–57.

Verna, C., A. Ramette, H. Wiklund, T. G. Dahlgren, A. G. Glover, F. Gaill, and N. Dubilier. 2010. High symbiont diversity in the bone-eating worm *Osedax mucofloris* from shallow whale-falls in the North Atlantic. Environmental Microbiology 12:2355–70.

Vollrath, F. 1998. Dwarf males. Trends in Ecology and Evolution 13:159–63.

Vollrath, F., and G. A. Parker. 1992. Sexual dimorphism and distorted sex ratios in spiders. Nature 360:156–59.

Vrijenhoek, R. C., S. B. Johnson, and G. W. Rouse. 2008. Bone-eating *Osedax* females and their "harems" of dwarf males are recruited from a common larval pool. Molecular Ecology 17:4535–44.

———. 2009. A remarkable diversity of bone-eating worms (*Osedax*; Siboglinidae; Annelida). BMC Biology 7:74. doi: 10.1186/1741-7007-7-74

Walter, A., and M. A. Elgar. 2012. The evolution of novel animal signals: Silk decorations as a model system. Biological Reviews 87:686–700.

Walter, A., M. A. Elgar, P. Bliss, and R.F.A. Moritz. 2008. Wrap attack activates web-decorating behavior in *Argiope* spiders. Behavioral Ecology 19:799–804.

Webber, N. H. 1977. Gastropoda: Prosobranchia. Pp. 1–97 *in* A. C. Giese and J. S. Pearse, eds. Reproduction of Marine Invertebrates. Volume IV. Molluscs: Gastropods and Cephalopods. Academic Press, New York.

Weckerly, F. W. 1998. Sexual size dimorphism: Influence of mass and mating systems in the most dimorphic mammals. Journal of Mammalogy 79:33–52.

Wedell, N., M.J.G. Gage, and G. A. Parker. 2002. Sperm competition, male prudence and sperm-limited females. Trends in Ecology and Evolution 17: 313–20.

Weiss, L. A., L. Pan, M. Abney, and C. Ober. 2006. The sex-specific genetic architecture of quantitative traits in humans. Nature Genetics 38:218–22.

Weiss, S. L. 2006. Female-specific color is a signal of quality in the striped plateau lizard (*Sceloporus virgatus*). Behavioral Ecology 17:726–32.

Wells, M. J., and J. Wells. 1977. Cephalopoda: Octopoda. Pp. 291–348 *in* A. C. Giese and J. S. Pearse, eds. Reproduction of Marine Invertebrates. Volume IV. Molluscs: Gastropods and Cephalopods. Academic Press, New York.

West-Eberhard, M. J. 1983. Sexual selection, social competition, and speciation. Quarterly Review of Biology 55:155–83.

Weston, E. M., A. E. Friday, and L. Pietro. 2007. Biometric evidence that sexual selection has shaped the hominin face. PLoS One 2:e710. doi: 10:1371/journal .pone.0000710.

White, F. 1969. Distribution of *Trypetesa lampas* (Cirripedia, Acrothoracica) in various gastropod shells. Marine Biology 4:333–39.

———. 1970. The chromosomes of *Trypetesa lampas* (Cirripedia, Acrothoracica). Marine Biology 5:29–34.

Wiens, J. J., T. W. Reeder, and A. N. Montes de Oca. 1999. Molecular phylogenetics and evolution of sexual dichromatism among populations of the yarrow's spiny lizard (*Sceloporus jarrovii*). Evolution 53:1884–97.

Wiklund, C. 2003. Sexual selection and the evolution of butterfly mating systems. Pp. 67–90 *in* C. L. Boggs, B. W. Watt, and P. R. Ehrlich, eds. Butterflies: Ecology and Evolution Taking Flight. University of Chicago Press, Chicago, London.

Wilder, S. M., and A. L. Rypstra. 2008. Sexual size dimorphism predicts the frequency of sexual cannibalism within and among species of spiders. The American Naturalist 172:431–40.

Wilder, S. M., A. L. Rypstra, and M. A. Elgar. 2009. The importance of ecological and phylogenetic conditions for the occurrence and frequency of sexual cannibalism. Annual Review of Ecology and Systematics 40:21–39.

Williams, J., A. Gallardo, and A. Murphy. 2011. Crustacean parasites associated with hermit crabs from the western Mediterranean Sea, with first documentation of egg predation by the burrowing barnacle *Trypetesa lampas* (Cirripedia: Acrothoracica: Trypetesidae). Integrative Zoology 6:13–27.

Williams, K. L. 2003. The relationship between cheliped color and body size in female *Callinectes sapidus* and its role in reproductive behavior. Masters Thesis, Zoology, Texas A & M University, Texas A & M University Libraries Digital Publication, College Station, Texas.

Wilson, H. R. 1991. Interrelationships of egg size, chick size, posthatching growth and hatchability. World's Poultry Science Journal 47:5–20.

Wood, J. B., and R. K. O'Dor. 2000. Do larger cephalopods live longer? Effects of temperature and phylogeny on interspecific comparisons of age and size at maturity. Marine Biology 136:91–99.

Woodroffe, R., and A. Vincent. 1994. Mother's little helpers: Patterns of male care in mammals. Trends in Ecology and Evolution 9:294–97.

Worsaae, K., and G. W. Rouse. 2009. The simplicity of males: Dwarf males of four species of *Osedax* (Siboglinidae; Annelida) investigated by confocal laser scanning microscopy. Journal of morphology 271:127–42.

Yamaguchi, S., Y. Ozaki, Y. Yusa, and S. Takahashi. 2007. Do tiny males grow up? Sperm competition and optimal resource allocation schedule of dwarf males of barnacles. Journal of Theoretical Biology 245:319–28.

Young, R. E. 1996. *Tremoctopus* eggs, embryos and hatchlings. http://tolweb .org/accessory/Tremoctopus_Eggs,_etc.?acc_id=2416 *in* The Tree of Life Web Project, http://tolweb.org/. Accessed August 27, 2011.

———. 2010. Alloposidae Verrill 1881. *Haliphron atlanticus* Steenstrup 1861. Version 15 in August 2010 (under construction). http://toweb.org/Haliphron _atlanticus/20200/2010.08.15 *in* The Tree of Life Web Project, http:// tolweb.org/. Accessed August 15, 2010.

Young, R. E., and M. Vecchione. 2008. Argonautoidea Naef. 1912. Version 21, October 2008. http://tolweb.org/Argonautoidea/20192/2008.10.21 *in* The Tree of Life Web Project, http://tolweb.org/. Accessed August 25, 2011.

Zeh, J., and D. W. Zeh. 2003. Toward a new sexual selection paradigm: Polyandry, conflict and incompatibility. Ethology 109:929–50.

Zuk, M. 1991. Sexual ornaments as animal signals. Trends in Ecology and Evolution 6:228–31.

———. 2002. Sexual Selections. What We Can and Can't Learn about Sex from Animals. University of California Press, Berkeley, California.

ILLUSTRATION CREDITS

Figure 7.1. Reprinted from Thomas, R. F. 1977. Systematics, distribution, and biology of cephalopods of the genus *Tremoctopus* (Octopoda: Tremoctopodidae). Bulletin of Marine Science 27(3):353–92, p. 366, figure 6d, with permission of the Bulletin of Marine Science.

Figure 8.1. Reprinted from p. 2 of Bertelsen, E., ed. 1951. The Ceratioid Anglerfishes. Ontogeny, Distribution and Biology. Bianco Luno, Copenhagen, with kind permission of the Zoological Museum, University of Copenhagen.

Figure 8.2. Reprinted from Regan, C. T. 1925. Dwarfed males parasitic on the females in oceanic angler-fishes (*Pediculati ceratioidea*). Proceedings of the Royal Society of London. Series B 97(684):386–400, figure 3, p. 390, with permission of the Royal Society of London.

Figure 9.2. Photographs kindly provided by Greg Rouse, Scripps Institution of Oceanography, University of California at San Diego, La Jolla, California.

Figure 10.1. Reprinted from Gotelli, N. J., and H. R. Spivey. 1992. Male parasitism and intrasexual competition in a burrowing barnacle. Oecologia 91:474–80, figure 1, p. 475, with permission from Springer Science and Business Media.

Plate 1. Photograph kindly provided by Derek Roff, Department of Biology, University of California, Riverside, California.

Plate 2. Photograph kindly provided by Tamás Székely, Department of Biology and Biochemistry, University of Bath, UK.

Plate 3. Photograph kindly provided by Derek Roff, Department of Biology, University of California, Riverside, California.

Plate 4. Photograph kindly provided by Derek Roff, Department of Biology, University of California, Riverside, California.

Plate 5. Photograph kindly provided by Derek Roff, Department of Biology, University of California, Riverside, California.

Plate 6. Photograph kindly provided by Franz Kovaks, Orth/Donau, Austria, www.kovacs-images.com.

Plate 7. Photograph kindly provided by Franz Kovaks, Orth/Donau, Austria, www.kovacs-images.com.

Plate 8. Photograph printed with permission of David Tipling, David Tipling Photography, Holt, Norfolk, UK, www.davidtipling.com.

Plate 9. Photograph kindly provided by David Kjaer, Warminster, Wiltshire, UK, www.davidkjaer.com.

Plate 10. Photograph kindly provided by David Kjaer, Warminster, Wiltshire, UK, www.davidkjaer.com.

Plate 11. Photograph kindly provided by Melchior W. N. de Bruin.

Plate 12. Photograph kindly provided by Ad Konings, Cichlid Press, El Paso, Texas, www.cichlidpress.com.

Plate 13. Photograph kindly provided by Troy Bartlett, Roswell, Georgia, www.naturecloseups.com.

Plate 14. Photograph printed with permission of Tom Murray, Groton, Massachusetts, www.pbase.com/tmurray74.

Plate 15. Photograph kindly provided by Cassandra L. LeMasurier, Coconut Creek, Florida.

Plate 16. Photograph from Norman, M. D., D. Paul, J. Finn, and T. Tregenza. 2002. First encounter with a live male blanket octopus, the world's most sexually size-dimorphic large animal. New Zealand Journal of Marine and Freshwater Research 36:733–36, figure 1a, p. 734, © The Royal Society of New Zealand. Reprinted by permission of the publisher (Taylor & Francis Ltd, http://www.tandf.co.uk/journals).

Plate 17. Photographs kindly provided by Greg Rouse, Scripps Institution of Oceanography, University of California at San Diego, La Jolla, California.

INDEX